AF689387

Building for Health
Andrew Pace

ISBN-13: 979-8-234-09054-6

The information presented in this book is intended solely for educational and informational purposes. The author shares personal opinions, research, observations, interviews, industry experience, and interpretations related to healthy homes, indoor air quality, building materials, environmental exposures, and wellness-oriented construction practices. This book is not intended to diagnose, treat, cure, or prevent any disease or medical condition, nor should it be considered medical advice. Readers experiencing health concerns, environmental illness, chemical sensitivities, respiratory symptoms, or other medical conditions should consult qualified healthcare professionals and licensed environmental specialists.

The author is not acting as a licensed physician, industrial hygienist, engineer, architect, contractor, attorney, code official, or environmental remediation professional unless explicitly stated otherwise. Readers should consult appropriately licensed and qualified professionals before making decisions regarding construction, remodeling, demolition, mold remediation, HVAC systems water quality, ventilation, product selection, environmental testing, or other building-related matters.

Many topics discussed in this book involve emerging research, evolving building science, environmental medicine, toxicology, and indoor environmental health concepts for which scientific understanding and industry standards continue to develop. Some recommendations or viewpoints may not reflect mainstream industry practices, regulatory positions, manufacturer guidance, or governmental standards.

Unless otherwise expressly stated, statements contained in this publication regarding products, materials, manufacturers, certifications, organizations, building practices, environmental exposures, health implications, or industry standards constitute the author's opinions, interpretations, analysis, and commentary based upon information available at the time of publication. Such statements are not intended as assertions of undisputed fact. Readers are encouraged to independently evaluate information and consult multiple sources when forming conclusions.

Environmental conditions and human responses vary widely. Individual sensitivities to mold, volatile organic compounds (VOCs), allergens, fragrances, chemicals, dust, electromagnetic fields, combustion byproducts, moisture, or building materials differ significantly from person to person. No guarantee is made that applying the concepts or recommendations in this book will produce specific health, environmental, or building performance results.

Construction practices, environmental conditions, moisture dynamics, ventilation requirements, and applicable building codes vary by climate, region, jurisdiction, and building type. Readers are responsible for verifying compliance with local laws, codes, regulations, and manufacturer specifications before implementing any methods discussed in this book.

The author and publisher expressly disclaim liability for any loss, injury, illness, adverse health effect, property damage, financial loss, construction defect, remediation outcome, code violation, or other consequence allegedly resulting from the use, misuse, interpretation, or application of information contained in this publication.

Readers are solely responsible for evaluating and implementing any information, recommendations, or suggestions contained in this book and assume all risks associated with their use.

Building for Health

Andrew Pace

July 31, 2026

For those who know my story, I'm a Healthy Home Concierge® and the founder of Green Design Center®, the oldest healthy home supply company in the US.

Now the long story....My family owns a commercial construction material supply company that dates back to the 1930's. Growing up, the dinner table conversations were not only about school and sports, they were about products, architects and contractors. So after I finished my education, it was an easy decision for me to step into the family business.

I joined the construction industry and had a desire to embark on new and exciting sales opportunities. With great interest, I learned all about the concrete waterproofing and coatings world and developed a market for my products throughout the Midwest. I was involved in just about every trade organization for commercial construction and architecture and developed friendships that live on. Well on the way to success in this industry, an event occurred that would change our company, and my life, forever. In 1992, I was awarded a contract to provide high-performance finishes for a large project high rise building in Milwaukee WI. Our crew just finished applying the first coat of a "low- VOC" water based epoxy, and the phone starts ringing. Folks living in the condos above were forced to evacuate due to a strong chemical aroma pushing up through the elevator shaft. Within minutes, our own crew members started having intense reactions, leading to an ambulance ride to the hospital due to severe breathing difficulties. Common knowledge at the time was that water based paints were safe. We learned a hard lesson that day.

One of my industry friends introduced me to a small niche company in San Diego that catered to people with "chemical sensitivities", a term completely foreign to me back then. American Formulating and Manufacturing (AFM) produced paints, stains, sealers and other specialty products that were specifically formulated to eliminate chemical off gassing and create a healthier space for the occupants. This was my epiphany. This was when I decided what my life mission would be.

For over a year, I attempted to promote these products in the same manner as our others. Architectural presentations were well received, but the commercial building industry wasn't ready for healthier options. This was 1992. The term Green Building was in its infant stages and architects were afraid to specify products that didn't contain the usual host of chemicals. I spent quite some time on sales calls to hospitals, nursing homes and schools; all places one might think could use these healthier materials. The administrators of these facilities did not believe in building related illness and therefore could not justify the small price increase for my products. However the doctors, nurses and teachers who approached me after the sales presentations were quite interested in purchasing for use in their own homes. Selling retail was foreign to our commercial company, but soon decided it was the way forward.

I formed a new retail business and filled up our mail order catalog (pre-internet) with AFM paints and stains, air and water purification devices, and a host of other personal care products, all designed to help those suffering from chemical sensitivities. In the following 10-15 years, the construction industry became focused on green building issues, such as energy efficiency, sustainability and global environmental concerns, but we maintained our human health first approach to take care of our customers.

Throughout the next two decades, we developed an extensive website, survived a housing market crash and pushed though an international pandemic. We sharpened our focus and clarified our message. Building For Health is not just a slogan, it's my mission. This book is the culmination of my 35 years of learning from those who have suffered with chemical sensitivities and have shared their experiences with me. I have worked with over 30,000 customers over the years and have learned from them all. By no means should this be used in lieu of medical advice, but please use the information contained here to make your own educated decisions about your home health.

Contents

Chapter 1

Introduction

A healthy home minimizes household sources of inflammation. This book is about designing, building, improving, maintaining, and furnishing a healthy home.

1.1 Inflammation and Chronic Illness

Chronic illness is on the rise. Approximately 6 out 10 of Americans have at least one chronic illness. Ninety percent of healthcare expenses in the United States go towards treatment and management of chronic illness.[2] Environmental toxins, both indoor and outdoor, contribute to almost all chronic illness. The majority of the chronic illness burden related to environmental toxins is borne by children.[5] This problem is finally getting the attention it deserves. The air we breathe is just as important as the food we eat and the water we drink. We invest in all of these things for the same core reason — to reduce bodily inflammation.

Inflammation in the body contributes to the development and expression of chronic illness. This is a foundational principle of functional medicine. Among medical schools of thought, functional medicine has been the most successful at addressing the root causes of chronic illness.[1]

Inflammation is a poorly understood term. Inflammation is merely a biological response to harmful stimuli.[3] Accordingly, everything harmful that causes a biological response is an inflammogen. A comprehensive list of inflammogens, if one could ever be compiled, would be both large and banal.

If you have bumped your shin on a coffee table, you have sustained a physical injury that caused a common form of inflammation known as a bruise. A bruise is a localized and acute form of inflammation caused by physical trauma to a muscle. The inflammation associated with the bruise subsides as the bruise heals. If you have lived in a house with mold, you have inhaled mold spores and mold-related toxins (both inflammogens) for a long period of time. Continuous exposure is more likely than acute injury to contribute to the development of chronic illness. Repeat exposure to specific inflammogens is likely to exacerbate sensitivities to similar and other inflammogens. In other words, repeatedly striking the same spot on your shin without fully allowing it to heal will weaken the shin and worsen the injury. Repeatedly inhaling mold-related inflammogens will increase your sensitivity to all mold-related inflammogens. Worse yet, a weakened shin is more susceptible to breaks and strains. Similarly, an individual suffering from continuous mold exposure will become more sensitive to other allergens and chemicals. These individuals may develop overlapping sensitivities that may ultimately be diagnosed as Multiple Chemical Sensitivity (MCS), Chronic Inflammatory Response Syndrome (CIRS), Mast Cell Activation Syndrome (MCAS), or Electromagnetic Hypersensitivity (EHS). This is unlikely a comprehensive list.

In this context, the term *sensitivity* describes the relationship between the quantity of the exposure and the severity of the corresponding inflammation. To increase your sensitivity to an inflammogen is to increase the inflammation produced by same amount of exposure to that inflammogen. We as laypeople understand sensitivity at an intuitive level through the mechanism of allergic reactions. For example, if you are bitten by a single ant, you may at first develop a small and mildly itchy pustule on your skin. If you are later bitten by a hundred ants at once, you may develop a severe reaction where your whole hand swells up like a balloon. After this severe reaction has subsided, you will likely have a much more severe reaction to singular bites. You may even diagnose yourself as *allergic* to ants. This notion of a severe allergy is a specific example of how we can develop increased sensitivity to a

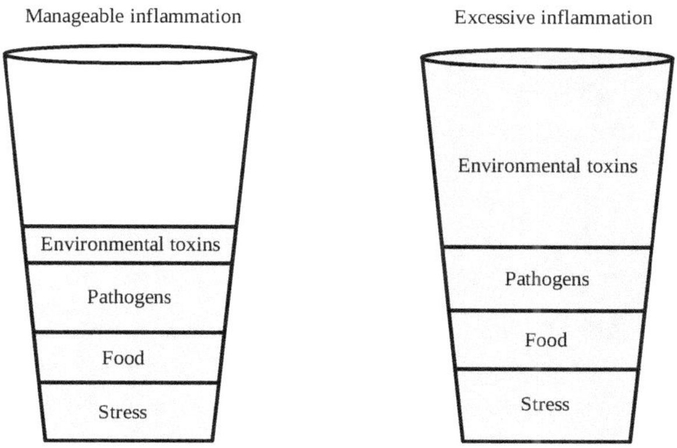

Figure 1.1: The Inflammation Bucket

particular inflammogen.

Functional medicine believes that chronic illness is a symptom of chronic inflammation. The theory is as follows. Our bodies have a total tolerance for inflammation from all sources. This tolerance corresponds to our body's total ability to successfully respond to harmful stimuli and its associated inflammation. If we exceed our total tolerance, we start to develop medical symptoms. Functional practitioners use the metaphor of a bucket to illustrate this principle.[4] The inflammation bucket metaphor asserts that any medical symptom associated with inflammation can be treated by reducing total inflammation. Therefore, if the removal of an inflammogen does not resolve a symptom, then the removal of other inflammogens are warranted until the bucket stops overflowing.

A connection exists between chronic illness and the lived environment that science and medicine are still trying to fully understand. Inflammation is the thread that connects them. Accordingly, I use the terminology of inflammation throughout this book to describe harm from the lived environment, though without particular fidelity to medical exactness.

1.2 The Journey to a Healthy Home

A healthy home minimizes household sources of inflammation. This book will give readers the information they need to design, build, improve, maintain, and furnish a healthy home. I will frequently distinguish between the pre-construction and post-construction phases of home ownership and give special care to describing the optimal courses of action for each, regardless of whether or not the home was constructed with these principles.

The metaphor of the inflammation bucket is a double-edged sword. On one hand, it is bleak. The inflammation bucket tells us that almost everything is an inflammogen. It tells us that we cannot directly measure each inflammogen's contribution to our total inflammation. This is a daunting proposition. On the other hand, the inflammation bucket assures us that our journey has an end. It tells us that our goal is not to permanently eliminate every conceivable inflammogen. Rather, our goal is to bring our total inflammation comfortably below the threshold at which we experience symptoms.

Eliminating sources of inflammation in our homes is complicated business. Typically, architectural interventions are more expensive and more risky than medical interventions. For example, if we seek to remediate moldy spray foam insulation in the walls of our house, we have to consider both the dollar costs of the remediation and the risks of disturbing and spreading the mold. Compare this with the medical intervention of quitting smoking. Both interventions are important to addressing total inflammation. But, the latter intervention costs nothing and presents no substantial risks.

This book will give readers the tools to navigate the complex decision-making process of planning and executing healthy home projects. This book is intentionally not an encyclopedia of potential material choices, as the plethora of materials available requires specific review from client to client, to ensure each individual's requirements are met. My goal with this book is to help readers prioritize projects with potentially significant health impacts without excessive financial costs or lifestyle disruptions. As of the time of writing, this is possible. Healthy home materials are much more accessible today than they were thirty years ago.

In my experience, healthy homes do not have to look or feel different from common homes. Healthy homes make sensible design choices with high health impact, without costing all that much more. While the average healthy material substitute costs slightly more than its common equivalent, most of the *healthy* upgrades improve longevity of the home and the occupants mental wellbeing, which will be discussed in later chapters.

Building a new custom home can be a daunting process. Improving a common home can be even more intimidating. Each chapter of this book will include an executive summary where I simplify the hard decisions. The rest of each chapter will be my justification and analysis that led to those decisions. Throughout, I will distinguish between guidance for new construction and existing construction. Very rarely is a problem in an existing construction completely unaddressable.

Chapter 2 describes my philosophy. Chapter 3 describes the tools and vocabulary I use to analyze materials questions. Subsequent chapters address materials and systems that are particularly relevant to occupant health.

References

[1] Jeffrey S. Bland. *The Disease Delusion: Conquering the Causes of Chronic Illness for a Healthier, Longer, and Happier Life*. Paperback. Foreword by Mark Hyman. HarperOne, 2014, p. 409. ISBN: 978-0062290748.

[2] NIHCM Foundation. "The Growing Burden of Chronic Diseases". In: (2025). URL: https://nihcm.org/publications/the-growing-burden-of-chronic-diseases.

[3] John M. Jennings, Saira Sabir, and Lauren Christie. "Pathology, Inflammation". In: *StatPearls*. 2024. URL: https://www.ncbi.nlm.nih.gov/books/NBK534820/ (visited on 11/03/2025).

[4] Sheila Kilbane and Tamela Rich. *Healthy Kids, Happy Moms: 7 Steps to Heal and Prevent Common Childhood Illnesses*. 2021.

[5] World Health Organization. "Chemical Safety Data and Statistics". In: (2017). URL: https://www.euro.who.int/en/health-topics/environment-and-health/chemical-safety/data-and-statistics.

Chapter 2

Principles for Risk Assessment

I must properly define *healthy* for the purposes of this book. Healthy building materials minimize inflammation for inhabitants of the home. In this context, *healthy* is a strict synonym for *non-inflammatory*. Since a building material can not be intrinsically health-promoting, a healthy building material is any building material that is not health-degrading.

To determine whether or not a building material is healthy, we must exercise sound principles of risk assessment. Homeowners must make risk assessments using their finite time, resources, and expertise. In theory, any one can research any thing for any amount of time. In reality, homeowners do not have enough time to develop cutting-edge expertise in every aspect of home design. This book aims to educate homeowners on specific materials while instilling in them a keen intuition for risk assessment. Accordingly, this book will adopt the perspective of a highly critical, risk-averse, and non-scientific homeowner. I call this perspective the *reasonable homeowner* perspective.

This chapter will explain and define the risk assessment framework we, as reasonable homeowners, use to study building materials. For this book to classify a building material as healthy, the reasonable homeowner must be convinced of its non-health-degrading nature using a mixture of the following principles.

2.1 Green Materials

Products that are marketed as *green* are not necessarily healthy. [The greenness of a product refers vaguely to its environmental impact, potentially along an unknown or disputed dimension. When products are green because their environmental impact has been engineered downward, their greenness bares no relationship to their healthiness.] When products are marketed as green solely on the basis of recyclability, they make no claim as to their healthiness. For example, recycled plastics are marketed as green despite the fact that the plastics recycling industry is highly toxic to human health and the environment.[1]

Greenness overlaps with healthiness only in a material's proximity to nature. Accordingly, many of the materials I recommend in this book will be green. Many of the vendors I recommend in this book rely on green marketing. Still, I will never rely on the purported greenness of a product in making a determination about its healthiness. I celebrate the right of homeowners to prioritize healthy materials to the detriment of all competing concerns.

Principle #1

Green products are not necessarily healthy products. Homeowners are morally justified in evaluating building materials solely on the basis of their healthiness.

2.2 Biological Mechanisms of Action

The developed world is highly susceptible to marketing arguments based on biological mechanisms of action (hereafter "BMOAs"). An argument based on a BMOA might possibly take one of the following forms.

- This product is healthy because chemical "X" reacts with chemical "Y" to produce chemical "Z" to alter environmental condition "W".

- This product is healthy because substance "X" interacts with the receptors of organelle "Y" to create beneficial effect on your body "Z" on which makes you feel emotion "W".

For example, I once had a well-meaning HVAC technician tell me that an ionizer would improve my indoor air quality. When I inquired as to how, he explained that the ionizer "just pumps electrons into the air to break down molecules". This is an example of a sales and marketing argument based on a BMOA. The technician's argument impliedly asserts that (1) chemically modifying the air in this way is always harmless and (2) larger molecules always become safer to breathe when broken down into smaller component molecules. A reasonable homeowner completely ignorant to the intricacies of chemistry and biology would find his statement overbroad and untrustworthy. In a later chapter, I will discuss exactly why. For now, it suffices to say that we as a culture are all susceptible to this sort of pseudo-scientific window dressing. We ought to be more wary of it.

Marketing arguments based on BMOAs are weak and dangerous because they assert the existence of a singular known BMOA without accounting for the innumerable other mechanisms by which that same action can affect our health. As a reasonable homeowner, I have no doubt that an ionizer can reduce bacteria in the air. At the same time, I know that we as humans will never fully comprehend all of other ways in which ionizers interact with all of the other chemicals in the air to affect our health.

Principle #2

The mere existence of a known harmful biological mechanism of action weighs heavily against a determination of safety for a material. The existence of a beneficial biological mechanism of action does not preclude the existence of other harmful mechanisms. A known harmful mechanism may be disregarded in a determination of safety only if its effects can be (1) mitigated effectively, (2) proven harmless under appropriate use, and (3) understood through common sense.

2.3 Adverse Substitution

Substances with known harmful BMOA's are often replaced by others with unknown harmful BMOA's through a process called adverse substitution. The nature of substance regulation in the U.S. almost guarantees this. Most of the U.S. laws that attempt to protect homeowners from harmful substances operate on a blacklist model. In a blacklist model, specific substances are only subjected to increased scrutiny after they are demonstrated or suspected of being harmful. In the U.S., the Toxic Substances Control Act (TSCA) (as amended in 2016) is one such example of a blacklist law regulating substances.

Blacklist laws endanger homeowners by encouraging adverse substitution. Adverse substitution occurs when a somewhat hazardous and well-researched substance is regulated out of usage only to be replaced by a more hazardous and less-researched substance. Adverse substitution can also manifest organically through evolving consumer preferences and clever marketing. The harm attributable to the new substance may go unnoticed because it is less acute and more chronic than the harm attributable to the old substance. The trend towards low-VOC paints is an example of adverse substitution I will discuss at length.

Contrast a blacklist law with a whitelist law. In a whitelist law, chemicals with unresearched human health effects must be demonstrated to be safe through extensive research before release into the marketplace. The European Union's REACH law is one such example of a whitelist law regulating chemicals. The EU's approach to chemicals regulation mirrors our approach to risk management by respecting the following principle.

Principle #3

Absence of evidence is not evidence of safety.

2.4 Proximity to Nature

Proximity to nature is our shield against adverse substitution. In the absence of complete information, natural materials are the lowest risk option. No material is completely natural in its finished form. Similarly, no material is very useful in its completely natural form. Still, we as reasonable homeowners can be assured of the safety of a building material if we can easily tell the story of how the raw material was extracted from the earth and turned into a finished product.

From this principle I inform vendors how to market materials to me. For example, us reasonable homeowners are not likely to be impressed by big automated assembly lines with multistep chemical treatment and finishing processes. Instead, we likely prefer to see a highly mechanical manufacturing process that finishes materials using heat and pressure. We are not Luddites. We do not require hand-formed straw bricks for our homes. But, we are suspicious of complexity.

Principle #4

Proximity to nature is evidence of safety. Accordingly, simple and transparent manufacturing processes are evidence of safety.

2.5 Inertness

Inertness is the backup plan when we can not achieve proximity to nature. *Inertness* describes the non-reactivity of a substance subject to certain environmental and chemical conditions. For example, consider a solid material that is chemically stable below 200 degrees Fahrenheit as long as it is not exposed to strong acids or bases. This material could be a type of concrete or a type of hard plastic. Either way, if it is rendered inert upon installation and stays inert during normal usage, it cannot harm us. As reasonable homeowners, we prefer natural materials, but merely inert materials may be sufficient.

Principle #5

Any inert material can be safe if its inertness is proven and maintained.

2.6 The Human Body as a Sensor

Ultimately, the human body is the best source of information regarding the how the human body reacts to a substance. The body does not need to know the chemical formula of a substance in order to know if it should react to it. While certain sensors are useful, the body itself is the only universal sensor.

Some individuals are more sensitive than others. Some have specific sensitivities. Others have highly generalized sensitivities. Accordingly, the lived experiences of chemically sensitive people are fantastic data. At The Green Design Center, I have spoken to thousands of chemically sensitive people. Their experiences informs my judgement.

Principle #6

The lived experiences of chemically sensitive individuals are the strongest evidence for or against the safety of a material.

2.7 Shifting the Burden of Proof

This book should appeal to those who have been harmed by inflammogens in their homes. By reading this book, you are deciding to shoulder some responsibility for preventing harm to yourself and your loved ones by developing expertise in selecting building materials. Economically, this is a tragedy. The materials manufacturers (often called "original equipment manufacturers" or "OEMs") are much better equipped to ensure the safety of their own products. OEMs are the only parties (to the exclusion of wholesalers, retailers, homeowners, and regulators) with the requisite knowledge and resources to adequately test their products. Accordingly, homeowners should feel comfortable requiring transparency from OEMs.

Principle #7

OEMs are the least-cost avoiders of harm from the materials they produce. OEMs are the best equipped organizations to (1) publish lists of ingredients, (2) describe manufacturing processes, and (3) commission independent testing of their products. Transparency from OEMs is worthy of trust from homeowners.

References

[1] Maria Doa and Sara Hull. *"Advanced recycling" is a toxic scam — now the EPA is turning a blind eye to some of the most toxic chemicals it produces.* Environmental Defense Fund. 2025. URL: https://blogs.edf.org/health/2025/08/04/advanced-recycling-is-a-toxic-scam-now-the-epa-is-turning-a-blind-eye-to-some-of-the-most-toxic-chemicals-it-produces/ (visited on 11/03/2025).

Chapter 3

Tools for Risk Assessment

This chapter will characterize the professional disciplines that affect the health of building materials. Each of these disciplines has something to contribute to my understanding of healthy building materials, but no single discipline has all the answers. With great respect for the efforts of these professions, I will detail the limitations of each. Research and data derived from each profession will inform my risk calculus for materials selection, which I will discuss in detail.

3.1 Biology

My second core principle of risk assessment describes the unbalanced utility of biological knowledge. (See Chapter 2, Principle #2, "The mere existence of a known harmful biological mechanism of action weighs heavily against a determination of safety for a material.") Here, I define biology to refer to generalized research into biological mechanisms of action (hereafter "BMOAs").

Biology as a science has a lot to be proud of. Consider that many of us learned about cellular respiration (and more specifically ATP synthesis) in high school biology class. Cellular respiration is the process by which the body turns food into energy. The discovery and description of the exact chemical formulae behind this process is a crowning achievement of modern science that beautifully marries humanity's collective knowledge of biology and chemistry. Cellular respiration is probably the most thoroughly researched BMOA.

Novel BMOAs are hard to discover. And, they are never really proven. For example, the theory of cellular respiration is considered virtually proven only because an enormous body of multidisciplinary research corroborates it from many different analytical angles. Other seemingly banal phenomena still lack a viable BMOA. For example, the BMOA for Fetal Alcohol Syndrome eludes biologists to this day.[7]

Biology only understands a small fraction of all conceivable BMOAs. There are millions of different chemicals that trigger different BMOAs according to their concentration and route of exposure. The complexity of the human body suggests that potentially millions of BMOAs exist. Biology has superficial knowledge of a small minority of all BMOAs. It has deep knowledge of only a few. Given this, how do we incorporate biological research into our risk calculus?

BMOAs can be categorized as (1) known or unknown, and (2) beneficial or harmful. Health-related product marketing in the United States will generally highlight known-beneficial BMOAs while disclosing known-harmful BMOAs. (See Chapter 2 for a discussion of the general form of BMOAs used in marketing.) Product marketing will necessarily exclude unknown-harmful BMOAs. In other words, no company will conclude an advertisement by stating, "What we don't know may harm you."

In the absence of alternative evidence, we ought to assume that the positive health effects of beneficial BMOAs are just as significant as negative effects of harmful BMOAs (hereafter the "effects"). Further, we ought to assume that unknown BMOAs outnumber known BMOAs by a factor of one hundred generally (hereafter the "prevalence"). Finally, in the absence of competing evidence, we will assume that each BMOA has an equally likely chance of personally affecting us (hereafter the "likelihood"). We can ascribe numerical estimates to each of these values to calculate the estimated positive or negative total effect of a product (hereafter the "risk value"). Throughout this section, each cell of the matrix ascribes a value to each type of BMOA. In the following tables, K indicates "known", U indicates "unknown", B indicates "beneficial", and H indicates "harmful". Each value is multiplied cell-wise and then added together to achieve a risk assessment. A positive number suggests a healthful decision and a negative number suggests a harmful decision.

The default calculation is described in Table 3.1. This calculation describes a theoretical action taken with no knowledge other than the existence of pleasure and pain coupled with an appreciation of the vast unknown. Its risk value is zero, suggesting neither action nor abstinence.

	Effect			Prevalence			Likelihood		
	B	H		B	H		B	H	
K	+1	−1	*	1	1	*	100%	100%	⇒ 0
U	+1	−1		100	100		100%	100%	

Table 3.1: The default risk assessment of a health-oriented product.

Given that a novel product is marketed on the basis of a beneficial BMOA, we may charitably assume that some benefits have at least been tested and recognized. Accordingly, we may prescribe a value of "+10" to the known-beneficial cell in my risk assessment. Further, we may assume that we will experience this known-beneficial effect with reasonable certainty, so I keep the corresponding prevalence and likelihood cells constant at "1" and "100%". Additionally, we may charitably assume that any obvious detrimental health effects were discovered during the testing and development of this product, so I reduce the likelihood of harmful effects (known and unknown) to "10%". Finally, we reasonably assume that the product was developed to target a specific beneficial BMOA, so the likelihood that we benefit from a unknown beneficial BMOA is low. Accordingly, I ascribe a value of "1%" to the unknown-beneficial likelihood cell. The risk value is described in Figure 3.2. This risk calculus might describe the installation of a reverse-osmosis water filter in a home.

	Effect B	Effect H		Prevalence B	Prevalence H		Likelihood B	Likelihood H	
K	+10	−1	*	1	1	*	100%	10%	⇒ +0.9
U	+1	−1		100	100		1%	10%	

Table 3.2: Risk assessment of a novel product marketed on the basis of a strong known-beneficial BMOA.

Many homeowners probably follow this thought process when deciding whether or not to purchase a water filtration system for their homes. The beneficial health effect of adding water filtration to a city tap water system is to remove some unhealthy chemicals from the water. The homeowner probably does not need to know much about the method of filtration or the nature of the unhealthy chemicals in the tap to be convinced to purchase the system. In any case, the filtered tap water system improves the homeowners risk posture against the default (i.e. the unfiltered tap water).

Now, consider the parameters of the previous example, but consider that you become aware of a known-harmful BMOA associated with the product. The known-harmful BMOA is not the primary intended effect of the product. Rather, it is a side effect. The side effect may not be all that severe. But, it exists, and you can be reasonably certain you will experience its detrimental effects over a long enough time using the product. To account for this knowledge, we increase the severity and likelihood of known-harmful effects (adjusting "-1" to "-2" and "10%" to "50%", respectively). Per this adjustment, the benefits of the product are eliminated. Note that Figure 3.3 renders a risk value of zero. Additional negative information will drive the risk value into negative territory.

Figure 3.3 might describe the risk calculus of installing an ionizer in your HVAC system. According to my current understanding, ionizers have a primary known-beneficial BMOA of neutralizing dust, but they have a secondary known-harmful BMOA of radicalizing existing indoor air pollutants. Arguments about the specific numerical values in the above risk equations are to be expected. This exercise is not scientific or exact. Readers are not encouraged to engage in it directly. Rather, this exercise demonstrates the sensitivity of health-related decisions to the presence of negative information.

	Effect			Prevalence			Likelihood		
	B	H		B	H		B	H	
K	+10	−2	*	1	1	*	100%	50%	⇒ 0
U	+1	−1		100	100		1%	10%	

Table 3.3: Risk assessment of a novel product with a strong primary known-beneficial BMOA and a weak secondary known-harmful BMOA.

Note that the prevalence matrix did not change throughout this discussion. The prevalence matrix represents the rate at which unknown BMOAs outnumber known BMOAs generally. Obviously, these values are debatable. Nonetheless, the factor of 100-to-1 used here reflects the following. We as reasonable homeowners are skeptical of humanity's ability to understand every conceivable BMOA. This skepticism weighs heavily on our risk calculus. My risk calculus is primarily a tug-of-war between the known-beneficial and the unknown-harmful. Weak known-harmful effects are frequent tie-breakers. Strong known-harmful effects immediately disqualify products from consideration.

This risk calculus is merely a demonstration. It is not a framework in which I am teaching you to follow. The underlying principles are easily absorbed. Pen-and-paper math is not needed to make these sorts of decisions. Rather, this mathematical example embeds a lot of the underlying truths about how we make decisions in the presence of incomplete information. Particularly, this demonstration elucidates how clever marketers can use known-beneficial BMOAs to influence our perception of risk. When we as homeowners respond with healthy skepticism, we reveal the unbalanced weight of biological knowledge. Negative biological knowledge weighs more heavily into my risk calculus than positive biological knowledge.

The forthcoming fields of science and law I will discuss are not ignorant to this reality. The field of toxicology has developed reference doses (RfD's) and reference concentrations (RfC's) to help account for uncertainty in risk calculations. Environmental law has adopted the Precautionary Principle for the same reasons. Industrial hygienists are more interested in protecting workers than understanding how chemicals hurt them. Industrial hygienists enlist epidemiologists to assist them in discovering patterns of harm among workers exposed to toxins. Both parties in turn collaborate with occupational safety regulators to attempt to keep workplaces safe. History is replete with stories of the successes and failures of each discipline. I am not concerned with these. Rather, we want to learn to speak their language so that we can leverage their work for our own risk assessments.

3.2 Toxicology

Toxicology is the scientific kernel of healthy home construction. It is the study of adverse reactions from exposure to toxins. Toxicology is distinguished from pharmacology by its focus on harm. Pharmacology is the study of beneficial BMOAs, while toxicology is the study of harmful exposure generally. Since building materials cannot be intrinsically health-promoting, we as homeowners focus on avoiding harm when selecting materials. Our priorities comport well with toxicology's focus on harm.

3.2.1 Dose-response data

To understand how toxicology defines risk, we must understand dose-response relationships. Specifically, we must understand how the data produced by dose-response experiments allow toxicologists to construct dose-response curves. A typical toxicology experiment tests the relationship between the following four factors: (1) a substance, (2) a subject, (3) a toxicity endpoint, and (4) an exposure strategy.

The substance can be any chemical, mixture, or product. The subject is generally a living organism that can react to an exposure (e.g. people, animals, bacteria, or organ cells). Some quantifiable amount of exposure forms the X-axis of the dose-response curve. A toxicity endpoint describes some manner in which a subject is expected to respond to an exposure. In human experiments, the skin might turn red. In animal experiments, cancer might develop. In cellular experiments, the cell might mutate or die. All of these responses are quantifiable toxicity endpoints. The

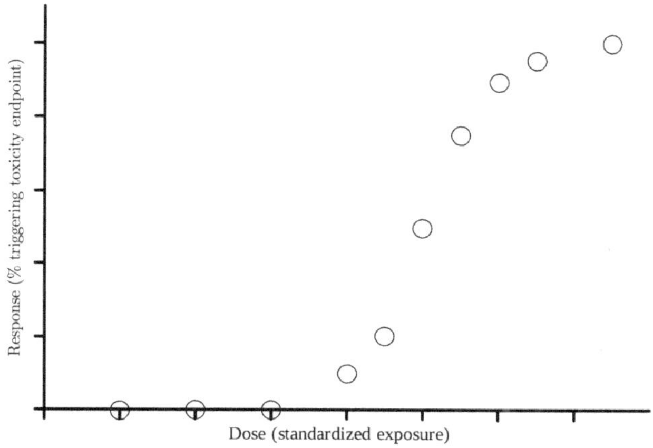

Figure 3.1: An example of a dose-response curve

aggregate response of a group of subjects forms the Y-axis of the dose-response curve.

The exposure strategy is the way in which the subject is exposed to the toxin. A common exposure strategy might be inhalation at a constant quantity for a fixed period of time. Another common exposure strategy might be ingestion of a fixed quantity every 24 hours.

For example, in an animal experiment, 100 mice might be split up into 10 groups of 10 and administered different doses of a solvent by inhalation. Each group is a point on the dose-response curve. The amount of exposure is the location of the point on the X-axis. The fraction of the group for which the toxicity endpoint is triggered is the location of the point on the Y-axis. The toxicity endpoints for this experiment might include tissue damage, acute intoxication, and mortality. In this example, we would ascertain three different dose-response curves, one for each toxicity endpoint.

The dose at which 50% of the mice in a group (i.e. 5 out of 10) trigger a toxicity endpoint is the effective dose (ED50) of the toxicity endpoint. The dose at which 50% of the mice die (i.e. the ED50

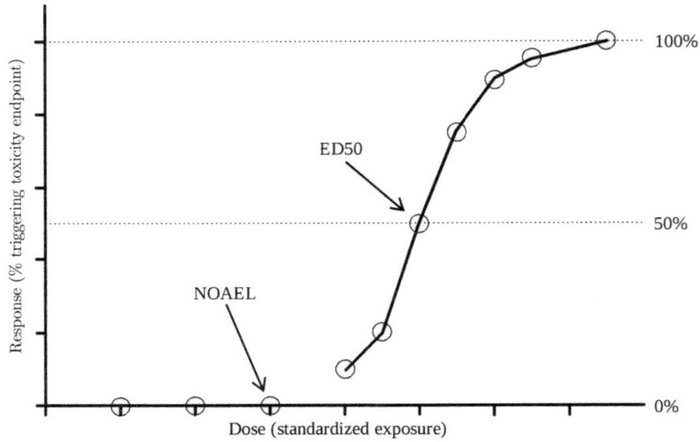

Figure 3.2: Identifying the ED50 and NOAEL on a dose-response curve

of the mortality endpoint) is the lethal dose (the LD50). The highest dose at which no biologically significant effect is observed is the No Observed Adverse Effect Level (NOAEL).

Ingestion exposure levels are often measured in dose-per-body-weight of the mouse, because the ability of an organism to effectively metabolize a toxin is typically proportional to its weight. Airborne toxins are often measured in airborne concentration (i.e. weight of substance in the air as a proportion of the weight of the toxified air). Airborne exposures do not account for subject weight because the subject weight is roughly proportional to the quantity of air the subject inhales. Accordingly, airborne concentrations automatically normalize for subject weight.

The astute reader may have noticed that the above-mentioned statistics (ED50, LD50, and NOAEL) rarely appear in laws and regulations. Laws and regulations contain a litany of acronyms to describe various actionable intakes and concentrations. Four of the most relevant figures are Reference Dose (RfD), Reference Concentration (RfC), Acceptable Daily Intake (ADI), and Personal Exposure Limit (PEL). These figures are set by policymakers using toxicological data. I will discuss one framework for determining an RfD from a NOAEL for illustrative purposes. Other values follow very similar methodologies.

To calculate an RfD, I'll take the NOAEL from an animal experiment and adjust it downwards to determine an exposure limit that can be comfortably applied to humans. The toxicological algorithm for calculating the RfD is beautiful in its risk-aversion. To calculate an RfD given a NOAEL, first determine if the animal testing data is strong or weak. If the animal testing data is weak (often due to a small sample size), then divide the NOAEL by 10. Then, divide that value by 10 to account for uncertainty in translating safe animal exposure to safe human exposure. Then, divide that value by 10 again to account for human variability and individual sensitivity. The result is the RfD. Thus, we have the following generalized rule for calculating an RfD based on the NOAEL from a small-scale animal study,

$$\text{RfD} = \frac{\text{NOAEL}}{1000}.$$

RfDs are the most commonly used benchmarks against which regulators set exposure limits. Statutory law (passed by the legislatures) typically gives broad authority to administrative agencies (e.g. the EPA and OSHA) to set substance-by-substance exposure limits. RfDs derived from toxicological studies are typically the source of these exposure limits. It can be easily argued that RfDs are too conservative. I believe they are perfectly conservative. Further, I believe the toxicological algorithm for setting RfDs reflects admirable intellectual honestly among toxicologists.

The consequences of this algorithm reflect a common sense approach to managing human exposure to toxins. For example, consider a chemical where the only available toxicological data is a small scale animal study that tested for cancer and tissue damage in mice. The RfD would be one one-thousandth of the highest exposure level at which no tissue damage or cancer was observed (i.e. the NOAEL divided by 1,000). Now, consider that a reasonable homeowner had to assess and potentially address his family's exposure to this chemical. The RfD would work as an actionable benchmark. If he were to take a measurement above the RfD, he would be spurred to action. He would have good reason to remediate the source of the exposure. While the NOAEL is hard data about cancer in mice, the RfD is a conservative estimate that reflects the personal values of risk-averse homeowners. The dose that causes cancer in lab mice is much greater than the dose that causes inflammation in infants. Naturally, the action level for humans is 1,000 times lower than the cancer-causing dose in mice.

3.2.2 The economics of toxicology

Toxicological studies are time-consuming and expensive. A recent analysis suggests that a two-generation reproductive animal study costs $350,000.[23] Other analyses suggest that pre-release testing of novel compounds sufficient to comply with European Union laws costs between $200,000 and $2,000,000.[14] The consequences of the economic reality are obvious. Toxicologists only have the resources to run comprehensive studies on (1) specific and common chemicals of concern and (2) novel chemicals with potentially high business value.

An important shortcut exists for certain well-researched chemicals of concern. Heavy metals and volatile solvents are illustrative examples. A heavy metal testing array (including common chemicals like lead and cadmium) can be purchased online and processed by a lab for $200. Similarly, government offices that support the farming industry will test soils for heavy metals, often for as low as $20. Similarly, volatile solvents (i.e. volatile organic compounds or "VOCs") can be field-tested with meters that cost approximately $400. Specific VOCs and their concentrations can be measured by a lab for a similar cost. While we as homeowners are grateful that common chemicals of concern can be tested cheaply, we are helpless to understand the specific concentrations and risks of unresearched and undisclosed chemicals in products we buy.

Toxicologists recognize their collective inability to test for adverse effects from all possible combinations of chemicals.[7] They also recognize that certain chemical combinations might synergistically exacerbate the toxicity of each other. Further, they recognize that testing for synergistic interactions between chemicals is economically intractable.[10]

Overall, toxicological data is very open and easy to access. The

National Institute of Health's open PubChem tool includes compilations of toxicological data from many sources and databases. Similarly, the Environmental Protection Agency's open CompTox tool compiles RfDs from many sources and databases. Still, the fraction of all economically significant chemicals for which sufficient toxicological data exists is trifling.[14]

3.2.3 Risk assessment with toxicological data

Like BMOAs in biology, toxicological data is a helpful tool for disqualifying well-researched hazardous materials. Consider how our risk calculus changes when selecting building materials as opposed to personal products. Many health-oriented products are designed to improve your health. Accordingly, their (1) known and unknown and (2) beneficial and harmful effects must be assessed. Materials (e.g. siding, insulation, and flooring) are not intrinsically health-promoting. They can only be health-degrading. The best material is one that poses no known or unknown risk of harm. Therefore, the risk values of materials are always negative or zero. Accordingly, the best material is the one with the least negative risk value. In my risk calculus for materials, I will ignore the possibility of beneficial health effects.

We also ought to consider that most materials do not intend to have any health effects at all. For example, drywall was not specifically designed to optimize inhabitant health. It is not necessarily unhealthy, and it may incorporate some health-conscious design choices, but it was certainly was not engineered specifically for health. The same can be said of pine boards used in framing or tiles used in floors. Accordingly, my default assumption will be that a building material has zero likelihood of impacting our health.

	Effect			Prevalence			Likelihood		
	B	H		B	H		B	H	
K	–	−1	*	–	1	*	–	0%	⇒ 0
U	–	−1		–	100		–	0%	

Table 3.4: The default risk assessment of a building material.

The most common challenge of healthy home building is selecting among different materials each with partially disclosed ingredients. Most commercially available materials only partially disclose their ingredients. Excluding materials specifically marketed with health and transparency in mind, manufacturers seem to disclose the minimum necessary information to comply with regulations affecting their materials. This places us in a difficult risk assessment predicament.

Common classes of proprietary ingredients in building materials include solvents, polymer precursors, and biocides. Commercially popular paints happen to use all three of these. Each of these classes of chemicals are well-known to be toxic through a variety of BMOAs. RfDs exist for well-known and well-researched members of these classes of chemicals.

The foundational principle of toxicology states that the dose, in consideration of individual sensitivity, makes the poison. Accordingly, a reasonable homeowner is open to using materials that contain potentially toxic chemicals if he can control and understand his potential exposure. In order to accurately understand the risk of using a specific material, the homeowner must be able to answer *yes* to the following questions, in this order.

1. **The transparency question.**
   Can I identify and quantify every ingredient in the material?

2. **The research question.**
   If so, does sufficient toxicological data exist for each ingredient?

3. **The safety question.**
   If so, will I be able to keep my exposure below the corresponding RfDs and RfCs during the installation phase and the use phase?

If the homeowner can answer *yes* to each of these questions, she can isolate her risk to the known-harmful. Further, the homeowner can be assured that any significant harm she may suffer from using the material would be caused (1) by professional negligence or accident on part of the installer or (2) by her failure to competently care for and maintain her home. Accordingly, in my risk assessment, I raise the likelihood of suffering known-harmful effects to the approximate likelihood of a failure to competently install and maintain, which I estimate at 2%. I maintain the likelihood of suffering unknown-harmful effects at 0% in accordance with the transparent and well-researched nature of the ingredients.

	Effect			Prevalence			Likelihood		
	B	H		B	H		B	H	
K	–	−1	*	–	1	*	–	2%	⇒ −0.02
U	–	−1		–	100		–	0%	

Table 3.5: The risk assessment of a transparent and toxicologically well-researched material, potentially containing hazardous chemicals, installed and maintained competently.

If the homeowner answers *no* to the transparency question, she cannot answer the research question or the safety question. If the homeowner answers *yes* to the transparency question but *no* to the research question, she again cannot answer the safety question. In either of these cases, the likelihood of suffering from an unknown-harmful effect rises significantly. Given the overwhelming general prevalence of unknown BMOAs, this weighs heavily against the material in my risk calculus.

When answering the safety question, the homeowner must consider basic questions about her environment and the intended use case for the material. For example, is the climate too hot or too cold for the material? Is the climate too dry or too humid? Will it be subjected to acute temperature changes? What about sunlight or other ultraviolet radiation? Will it be subjected to regular wear-and-tear? How so? Will it stand up to common spills and cleaners? How will it interact with acids, bases, alcohols, ammonia, and degreasers? Will it be easy feedstock for mold? If so, under what conditions? Under what circumstances do the chemicals in the material break down? Will the broken down chemicals remain airborne, settle on surfaces, or collect in dust?

Unknown-harmful effects from materials can arise (1) directly from hazardous ingredients in the materials and (2) indirectly from hazardous installation and use. Without access to ingredients lists and complete toxicological research, the homeowner cannot answer the safety question. This increases the likelihood of suffering an unknown-harmful effect. I ascribe a value of 20% to the unknown-harmful likelihood cell to reflect the lifetime possibility of suffering such an effect in the presence of such limited information.

	Effect			Prevalence			Likelihood			
	B	H		B	H		B	H		
K	–	−1	*	–	1	*	–	2%	⇒	−20.02
U	–	−1		–	100		–	20%		

Table 3.6: The risk assessment of a material with unknown ingredients or incomplete toxicological data.

For sake of completeness, I will discuss the risk assessment of a material for which the answer to the safety question is *no*. This discussion will elucidate how my desire for complete information stands at odds with the business objectives of materials manufacturers. Consider a material with completely transparent ingredients for which thorough toxicological data exists. Now, consider that the data suggests that this material is unsafe in all conceivable use cases. In my risk calculus, I would ascribe a value of 100% for the likelihood of suffering a known-harmful effect. Further, this known-harmful effect would likely be obvious and significant, so the magnitude of the negative health effect would carry a value of "-10". Further, I would adjust the likelihood of suffering an unknown-harmful effect to "20%" to reflect that the material in so broadly toxic that it is likely associated with exotic or very long-term health effects that have yet to be discovered. These values can be debated and would certainly vary from material to material. In any case, a finding of no possible safe usage is always immediately disqualifying for a material.

	Effect			Prevalence			Likelihood			
	B	H		B	H		B	H		
K	–	−10	*	–	1	*	–	100%	⇒	−30
U	–	−1		–	100		–	20%		

Table 3.7: The risk assessment of a transparent and toxicologically well-researched material for which no safe use case exists.

While this last point may same moot, it is not. Environmental law scholars have argued that the chemicals industry would need to spend only 0.2% of its annual revenue to develop complete toxicological data for all of its materials.[14] The authors of this study issued this estimate in an attempt to encourage corporations and regulators to make this practice the status quo. While the activism on part of the authors is admirable, the chemicals industry does not see the situation the same way. In reality, if the chemicals companies were to invest in developing this toxicological data, a significant fraction of all of their products would be immediately disqualified from use under the risk calculus of Table 3.7. Large chemical companies already operate on razor thin margins. Each disqualified product would vaporize an existing revenue stream and obviate significant prior capital expenditures. Such an initiative would threaten to bankrupt the chemicals industry.

3.3 Laws Affecting Building Materials

Laws affecting building materials help us in two ways. First, they attempt to keep homeowners safe from harmful substances in the materials they buy. Second, and more importantly, they create a complex system of disclosure requirements for manufacturers. These disclosure requirements provide important data to us as homeowners that we can use to inform our risk calculus. The laws themselves might be a weak patchwork, but we can still use what information they manage to surface. This chapter will discuss U.S. law for a U.S. audience.

I will start with a refresher on the basic principles of administrative law. Almost all laws affecting the usage of novel substances are administrative law (as opposed to common law, civil law, or criminal law). Administrative law is characterized by its use of enabling statutes that create administrative agencies with broad authority to accomplish some legislatively defined goal. Administrative laws have a common pattern. Generally, Congress (the legislative branch) passes a law that creates an agency to regulate a concept. The agency (now a new office of the executive branch) is given broad authority to study and regulate that concept The agency continually studies the concept and passes concrete regulations on the subject using its knowledge. The courts (generally the federal judicial branch) can hear and decide challenges to the practices of the agencies.

Accordingly, I will discuss many enabling statutes that task a handful of administrative agencies with various duties. The relevant agencies are the Environmental Protection Agency (EPA), the Occupational Safety and Health Administration (OSHA), the United States Department of Agriculture (USDA), and the Food and Drug Administration (FDA). Each enabling statute I will discuss tasks a specific one of these four agencies with the duty to regulate a specific concept in some way. Additionally, I will discuss the character and culture of each agency as a way to understand how each agency executes on its statutory responsibilities. The ultimate success of a statute seems to be greatly influenced by the priorities of it administrative stewards. I will discuss the statutes in roughly increasing order of importance to us as homeowners. Many of the statutes seem relevant but are not in reality.

3.3.1 Federal Insecticide, Fungicide, and Rodenticide Act (FIFRA)

First, the Federal Insecticide, Fungicide, and Rodenticide Act (FIFRA) authorizes the EPA to regulate the use of pesticides to prevent unreasonable adverse effects on the environment. The statute creates disclosure requirements for manufacturers of pesticides. Further, it gives special consideration to the impact of pesticides on the environment. Agriculture and urban sanitation employ an enormous volume of pesticides that can easily overwhelm and destroy natural ecology. Given this, the environmental posture of FIFRA makes sense. Additionally, FIFRA ties into the Federal Food, Drug, and Cosmetic Act (FFDCA) that is administered by the FDA. The FFDCA allows the EPA to regulate pesticide residues on food.[14]

Generally, FIFRA is focused on the environment and food. It generally exercises its authority on the factory floor, at the point of application, and at the point of sale. We as homeowners are primarily concerned about long-term exposure to pesticides in our indoor air and household dust.[13] Pesticides can be introduced to our indoor air when pesticides in our building materials offgas. And, pesticides can be introduced into our household dust from the soil surrounding our homes. The EPA acting under authorizations of FIFRA seems unlikely directly address these two long-standing sources of household pesticide exposure.

3.3.2 The Clean Air Act

The Clean Air Act seeks to protect the quality of ambient air (i.e. all indoor and outdoor air) through emissions regulations on stationary and mobile sources of air pollution.[14,17] The Clean Air Act is known as the smokestacks and tailpipes law. In the language of the Clean Air Act, *stationary sources* generally refer to factories and *mobile sources* generally refer to automobiles. The Clean Air Act was responsible for removing lead additives from gasoline and eliminating acid rain precursors from factory exhaust.[14] Both of these were remarkable successes for human health and the environment. Environmental law scholars claim that the Clean Air Act is the envy of international community, especially among countries that struggle with urban ambient air quality.[14] The Clean Air Act is broad enough, powerful enough, and flexible enough that the EPA was able to assume responsibility for greenhouse gas emissions under the auspices of stemming global warming without an act of Congress. (The Supreme Court ruled in *Massachusetts v. EPA (2007)* that greenhouse gases qualify as air pollutants under the Clean Air Act and that the EPA has authority to regulate them.[16])

We as homeowners need to understand what the Clean Air Act is in order to understand what it is not. A legal scholar might be able to intelligently argue that the Clean Air Act can protect homeowners from indoor air pollution from building materials. Regardless, history and precedent tells us that a different statute covers building materials and their effect on indoor air quality. That statute is a Toxic Substance Control Act.

3.3.3 Safety and the TSCA

The Toxic Substance Control Act (TSCA) gives the EPA authority to regulate virtually all chemical substances and mixtures that are not directly addressed by other administrative laws. Some products fit nicely into the mandate of the TSCA, like paints and insulations. Other products are exempt from the TSCA because they are preempted by other more specific laws. These products include foods, drugs, cosmetics, and pesticides.[22] Other concepts are more directly regulated by other statutes so will be handled by those laws for precedential reasons. For example, lead from automobile exhaust and sulfur dioxide from smokestacks will always be Clean Air Act issues despite that fact that each are toxic substances that can impact indoor air quality.[14]

The EPA is ineffective at promoting safety under the authority granted to it by the TSCA. The 2016 amendments to the TSCA required that the EPA identify, investigate, and regulate priority chemicals of concern.[14] The 2016 amendments gave the EPA broad discretion and authority to do so. Yet, the EPA took three years to plan and an additional five years to develop a risk evaluation on formaldehyde, finally releasing its formal guidance in December 2024.[21] During this same time period, a flurry of laws and regulations attempted to address human formaldehyde exposure more directly. A 2010 act of Congress specifically targeted formaldehyde (the Formaldehyde Standards for Composite Wood Products Act). This was followed by new EPA regulations in 2013, followed by a final rule from the EPA in 2016, followed by technical arguments through 2019. Finally, in 2019, the EPA finalized mandatory emissions standards for formaldehyde in wood products.[19] The standards are called "TSCA Title VI" and "CARB ATCM Phase II". U.S. sellers of engineered wood products were given until 2019 to come into compliance. An exception was made for domestic producers of laminated engineered wood products. These producers were given until 2024 to come into compliance.[5]

And, during that same time period, millions of people suffered from inflammation due to exposure to formaldehyde from building materials. A class action suit against a major materials distributor identified 760,000 people affected by flooring with excessive formaldehyde emissions from 2009 until the date the lawsuit was filed in 2015.[9] In all likelihood, the majority of those affected are unaware and have yet to replace their flooring. A class action suit against the federal government identified that emergency housing units provided to victims of Hurricane Katrina were emitting excessive formaldehyde and harming occupants.[8] This case confirmed that, despite the well-known risks of off-gassing from engineered woods, cheap materials were still polluting indoor air with dangerous concentrations of formaldehyde. Urea-formaldehyde resins are a direct manufacturing input into chipboards, particleboards, fiberboards, and pressboards, yet these materials are only still being incrementally phased out according to the abovementioned regulatory schedule.

Regulating formaldehyde should not be so complicated. Formaldehyde is an extremely well-understood hazardous chemical that has been used in resins of engineered would products for over one hundred years. Formaldehyde is an end-product of the breakdown of thousands of different chemicals, many of which appear in building materials. Formaldehyde's inflammatory health effects have been acknowledged as early as 1977. Thousands of exposure studies have been performed on formaldehyde.[15]

Formaldehyde itself is simple. It is the simplest aldehyde. Its chemical formula is CH_2O. To an organic chemistry student, it is barely more complex than water.[11] Formaldehyde and other complex aldehydes are extremely common in materials manufacturing. They are important ingredients in resins, polymers, adhesives, solvents, and biocides.[2,12,20] Accordingly, virtually all these types of products are liable to break down into formaldehyde when exposed to ultraviolet light, ozone, heat, or moisture.[6]

The story of formaldehyde is a specific example of the general failure of the EPA to protect consumers from hazardous chemicals in building materials under the TSCA. The TSCA Title VI and CARB ATCM Phase II requirements are relatively thorough. These regulations will probably succeed in preventing most adverse effects from formaldehyde exposure in building materials manufactured in 2025 and beyond. Unfortunately, we as homeowners will only realize the benefits of this regulation nearly 50 years after the EPA became aware of the threat posed by formaldehyde exposure. Further, the testing procedures specified under these regulations fail to account for certain common household use cases. We as homeowners still have to be proactively cautious about how we use engineered woods in our homes. (See Chapter 5 on engineered woods.)

Among all the chemicals in the world, formaldehyde should be one of the easiest to regulate. Formaldehyde is (1) very common, (2) not proprietary, (3) easy to understand, (4) easy to detect, (5) obviously harmful, and (6) completely non-essential. The EPA took 50 years to fulfill its duty to protect the public from this very simple compound. We as homeowners should have no confidence that the EPA will protect us from the thousands of other compounds commonly found in building materials, especially when some of the the most biologically significant compounds remain undisclosed to the public.

3.3.4 Transparency and the TSCA

The EPA is ineffective at promoting transparency under the authority granted to it by the TSCA. The TSCA, as implemented, contains a mechanism whereby chemical manufacturers can request that the EPA refrain from publicly disclosing the chemical formulas of ingredients in their products.[18] This tool is called a confidential business information ("CBI") claim. Under the TSCA, materials manufactures have to at least notify the EPA of new chemicals it will be releasing into the marketplace through a premanufacturing notice (PMN). Within a PMN, a manufacturer must state all known safety and hazard information about the ingredients in the product. A CBI claim allows a manufacturer to publicly censor almost any information in a PMN.

The TSCA gives the EPA broad discretion in how it responds to CBI claims. In the three and a half years following the enactment of the 2016 amendments to the TSCA, the EPA wholly rejected only 2.2% of all CBI claims (among 667 total claims). Among those 97.8% that it accepted, it used valueless legal boilerplate in its public justifications.[3] A cursory scan of the Federal Register reveals liberal use of CBI claims.[4] The legal tactics employed to achieve these redactions are both numerous and banal. It suffices to say that this liberal issuance of CBIs thwarts any meaningful toxicological research into these novel chemicals. At best, the scientific community might surface some product-specific epidemiological data. This too is a tragedy. Strong epidemiological data requires a lot of human suffering.

Clever marketers at chemicals manufacturers have even turned the new PMN requirement on its head by openly promoting the EPA's awareness of the compounds in their products. For example, some paints now advertise that they contain "EPA-registered mildewcides". The term *EPA-registered* suggests that the product is safe because it has been subjected to regulatory oversight. In reality, the term *EPA-registered* merely indicates that the mandatory PMN for the paint was filed. This fact provides us as homeowners no comfort. In all likelihood, the chemical formula for the advertised mildewcide is redacted by a CBI claim somewhere inside that PMN.

As reasonable homeowners selecting materials, we rarely need to engage directly with chemistry, biology, and toxicology. Most materials will simply fail on the transparency question because they intend to. The EPAs failure to advance ingredient transparency is often to blame.

3.3.5 Structural Deficiencies of the TSCA

I want to conclude this chapter by offering a path forward for chemicals regulation. I believe chemical manufacturers should bear the initial burden of proof of safety for novel substances.

The TSCA is fundamentally weak because it is a blacklist law. A blacklist law puts the burden of proof of safety on regulators. In such a law, the EPA must individually investigate and ban each problematic substance. The EPA will never keep up with the pace of new substances hitting the market. For example, the EPAs final ruling on formaldehyde contained over 2,000 pages of scientific modeling and research.[21] Charitably, we can say the EPA did thorough research. Realistically, it wasted tax dollars and human capital reevaluating the safety of one of the most well-researched indoor air pollutants in the world. While it is easy to deride government waste, I believe that even the most efficient regulatory agency would fail to achieve reasonable public safety under a blacklist regime.

The manufacturer is the least-cost avoider of harm from its products. As the inventor of the product, the manufacturer has near-perfect knowledge of the ingredients as well as any effects arising from common use and handling. If any party is going to intelligently and efficiently assess the safety of a novel chemical in its products, it will be the manufacturer. If the manufacturer can achieve such an assessment and pass a process audit of its findings, it should be allowed to introduce the chemical to the marketplace on the condition that it can provide full transparency as to the ingredients. The European Union's chemicals law (EU REACH) roughly corresponds to this approach. EU REACH is a whitelist law. Legal scholars call it a *no data, no market* law. Only chemicals on the whitelist are allowed to be sold.

The enabling statutes for these agencies prescribe this behavior. While the EPA sends scientists on dramatic investigations to build its blacklist, the EU REACH administration reviews neat little packets of toxicological data to add to its whitelist. There are too many novel chemicals on the market for the EPA to succeed in its mission to protect us from them. Alternatively, in the EU, there are only as many chemicals available on the market as there is time to authorize them.

Free market purists will decry the delay and expense hoisted onto industry by laws like EU REACH. They will assert that the restrictive laws in the EU are causing it to be left behind the rest of the world in terms of chemicals innovation. We as reasonable homeowners would like to see human health prioritized over chemicals innovation. There already exist safe alternatives to every toxic building material. I would prefer to see government and industry focus its collective attention on increasing the accessibility of safe materials and diminishing our reliance on toxic materials. The burgeoning human cost of chronic illness demands it.

3.4 Other relevant fields of study

There exist many specializations adjacent to toxicology and law that are relevant to healthy home design. A proper treatment of each of them is out of scope for this book, but readers should have an awareness of each. Each field has something important to say about materials. I will draw selectively from each throughout the remaining chapters.

3.4.1 Exposure Analysis

Exposure analysis is a field of applied toxicology that attempts to understand how toxins are ingested and exhausted from the human body. Exposure analysis is more likely to focus on large particles than pure chemicals.

An experiment in exposure analysis might seek to answer the following question. If a subject is placed in a water bath containing a certain concentration of a solvent, what concentration of the solvent can be measured in its exhaled breath? Exposure analysts might be interested in how the concentration of the solvent in the exhaled breath dissipates over time. This data might reveal information about the speed at which the body metabolizes and excretes the solvent through various biological pathways.

We have the field of exposure analysis to thank for our detailed understanding of how children can ingest heavy metals and pesticides from household dust. Exposure analysis often cuts to heart of problems that befuddle biologists and toxicologists in a way that is directly actionable for homeowners. For example, an exposure analyst may be able to accurately measure the pesticide exposure a child achieves through skin contact while playing outside. A parent does not require proof that the specific pesticide is harmful in order to take action to eliminate the child's exposure. Both the parent and exposure analyst are more than comfortable assuming that the exposure is dangerous.

While a homeowner might take his shoes off at front door to prevent the spread of germs, an exposure analyst does the same to prevent the accumulation of heavy metals in dust. Overall, exposure analysis touches many of the same areas of life as toxicology, but its findings are more actionable for individuals. It has a specific concern with indoor environmental health that we as homeowners share.

3.4.2 Industrial Hygiene

Industrial hygienists are practitioners responsible for ensuring workplace safety. They have battlefield knowledge. Industrial hygienists must engineer workplace environments to achieve an acceptable level of exposure in situations where some exposure in inevitable.

Industrial hygienists are largely responsible for ensuring compliance with toxin-related regulations promulgated by the Occupational Safety and Healthy Administration ("OSHA"). A lot of the frontline data about the dangers of specific chemicals comes from industrial hygienists. For example, we have industrial hygienists to thank for our thorough understanding of isocyanate toxicity. The general public has been saved from a good deal of harm from isocyanate exposure because factory workers manufacturing isocyanate products had the largest exposures and the worst symptoms.

3.4.3 Epidemiology

Epidemiologists are the boots-on-the-ground detectives of the health sciences. They have a close relationship with the public health establishment and public health policymakers. Long before RfDs and BMOAs are discovered, epidemiologists discover and investigate concerning patterns of illness in the public.

Epidemiology is our first line of defense, albeit a weak one, for understanding emerging harms. We can thank epidemiology for our understanding of the prevalence and evolution of chronic illness. As of the time of writing, we are still in the emergent phase of our understanding of chronic illness and inflammogens. Epidemiology has identified a problem, but the other branches of science, law, and policy have not addressed it.

Epidemiological studies are notoriously controversial. This is not due to some sort of bias, necessarily. Rather, epidemiologists have very limited access to data. Further, the data that they can acquire is often expensive or low-quality. This is merely the nature of the field.

If an epidemiologist believed peanut butter caused cancer, how would she prove this? Well, she would have to commission a survey to thousands of people about peanut butter consumption and cancer diagnoses. Then, she would have to make sure she controlled for the fact that peanut butter consumption might be correlated to smoking (and any other causes of cancer). The work quickly gets complicated, and the insights quickly get weaker and weaker.

Still, epidemiologists are the best equipped to identify mysterious and emerging issues. For example, it was probably an epidemiologist that discovered that cheaply manufactured mobile homes were causing formaldehyde exposure for thousands of people.

3.4.4 Molecular Toxicology

Molecular toxicology is a branch of toxicology that focuses on complex BMOA-driven problems. We have molecular toxicology to thank for our understanding of the dangers of Bisphenol A

("BPA") exposure at the parts-per-trillion level. BPA is a chemical commonly found in plastic food-grade containers that is an endocrine disruptor at single-digit parts per trillion concentrations.[1] Traditional toxicological methods were not well suited to discover the endocrine-disrupting potential of BPA. Understanding BPA required special methods introduced by molecular toxicology. BPA toxicity is distinct because it operates on a non-traditional toxicity endpoint at extremely low concentrations to produce primarily chronic symptoms.

Molecular toxicity teaches us that certain chemicals can be dangerous not because they cause tissue damage but because they mimic other functional chemicals in the human body. The description of BPA toxicity in humans is a great achievement in the field of biology. Yet, the discovery of BPA toxicity magnifies the weight of the unknown.

3.4.5 Certifiers

There exist many many organizations that attempt to certify building materials using various methodologies. Usually, the certifying organizations are interested in assuring consumers that the material is some combination of performance, green, or healthy. On average, these certifications are of little value to homeowners. They are simply too numerous, complex, and fragmented to be trusted. A homeowner could read this book cover-to-cover faster than she could understand the meaning behind all of the certifications used in just the carpeting industry.

In general, I am an advocate of The Declare List. The Declare List does not attempt to certify materials. Rather, it allows manufacturers to list the manufacturing inputs of their products as though they were foods on a grocery store shelf. The Declare List, per my knowledge, is opt-in, non-exclusive, and self-reported. A company can list only as many products as it would like to according to its sole discretion. And, the manufacturers self-reported list of manufacturing inputs is not checked for factual accuracy. Still, we as reasonable homeowners reward transparency from manufacturers, so I use The Declare List frequently.

References

[1] Jeffrey S. Bland. *The Disease Delusion: Conquering the Causes of Chronic Illness for a Healthier, Longer, and Happier Life.* Paperback. Foreword by Mark Hyman. HarperOne, 2014, p. 409. ISBN: 978-0062290748.

[2] William D. Jr. Callister and David G. Rethwisch. *Callister's Materials Science and Engineering.* Tenth. 2020.

[3] Richard Denison. *EPA still has a very long way to go on transparency under TSCA.* Accessed: 2025-12-23. Environmental Defense Fund. 2020. URL: https://blogs.edf.org/health/2020/01/28/epa-still-has-a-very-long-way-to-go-on-transparency-under-tsca/.

[4] Environmental Protection Agency. *Certain New Chemicals; Receipt and Status Information for May 2025.* 90 FR 40355; EPA-HQ-OPPT-2025-0067; FR Doc. 2025-15724. 2025. URL: https://www.federalregister.gov/documents/2025/08/19/2025-15724/certain-new-chemicals-receipt-and-status-information-for-may-2025 (visited on 12/23/2025).

[5] Environmental Protection Agency. *Compliance Date Extension; Formaldehyde Emission Standards for Composite Wood Products.* Federal Register. Final rule; compliance date extension. 40 CFR Part 770; EPA-HQ-OPPT-2017-0244; FRL-9966-56; RIN 2070-AK35. Sept. 2017. URL: https://www.regulations.gov/document/EPA-HQ-OPPT-2017-0244-0013.

[6] Charles R. Frihart, Timothy L. Chaffee, and James M. Wescott. "Long-Term Formaldehyde Emission Potential from UF- and NAF-Bonded Particleboards". In: *Polymers* 12.8 (2020), p. 1852. DOI: 10.3390/polym12081852.

[7] Steven G. Gilbert. *A Small Dose of Toxicology: The Health Effects of Common Chemicals.* Informa Healthcare, 2008.

[8] *In re FEMA Trailer Formaldehyde Products Liability Litigation.* 668 F.3d 281 (5th Cir. 2012). 2012.

[9] *In re Lumber Liquidators Chinese-Manufactured Flooring Products Marketing.* 2020 U.S. Dist. LEXIS 181103. 2020.

[10] Curis D. Klaassen. *Toxicology: The Basic Science of Poisons.* 9th. McGraw Hill Education, 2019.

[11] David Klein. *Organic Chemistry as a Second Language: Second Semester Topics*. Sixth. Wiley, 2024.

[12] Georgios M. Kontogeorgis and Kiil Søren. *Introduction to Applied Colloid and Surface Chemistry*. Wiley, 2016.

[13] Wayne R. Ott, Anne C. Steinemann, and Lance A. Wallace. *Exposure Analysis*. CRC Press, 2007.

[14] Percival et al. *Environmental Regulation*. Tenth. 2024.

[15] Tunga Salthammer, Sibel Mentese, and Rainer Marutzky. "Formaldehyde in the Indoor Environment". In: *Chemical Reviews* 110.4 (Jan. 2010), pp. 2536–2572. ISSN: 1520-6890. DOI: 10.1021/cr800399g.

[16] Supreme Court of the United States. *Massachusetts v. Environmental Protection Agency*. 549 U.S. 497. 2007.

[17] United States Environmental Protection Agency. *Summary of the Clean Air Act*. https://www.epa.gov/laws-regulations/summary-clean-air-act. 42 U.S.C. §7401 et seq. (1970). 2025. (Visited on 11/03/2025).

[18] U.S. Environmental Protection Agency. *40 CFR Part 703—Confidentiality Claims*. Title 40: Protection of Environment, Chapter I, Subchapter R, Part 703. Current through amendments as of December 19, 2025. 2025. URL: https://www.ecfr.gov/current/title-40/chapter-I/subchapter-R/part-703 (visited on 12/23/2025).

[19] U.S. Environmental Protection Agency. *Formaldehyde Emission Standards for Composite Wood Products*. https://www.epa.gov/formaldehyde/formaldehyde-emission-standards-composite-wood-products. Accessed: November 3, 2025. 2024.

[20] U.S. Environmental Protection Agency. *Laws and Regulations Concerning Formaldehyde*. https://www.epa.gov/formaldehyde/laws-and-regulations-concerning-formaldehyde. Accessed: 2025-11-03. 2025.

[21] U.S. Environmental Protection Agency. *Risk Evaluation for Formaldehyde*. Accessed: 2025-11-03. 2024. URL: https://www.epa.gov/assessing-and-managing-chemicals-under-tsca/risk-evaluation-formaldehyde.

[22] U.S. Environmental Protection Agency. *Summary of the Toxic Substances Control Act.* Accessed: 2025-11-03. 2016. URL: https://www.epa.gov/laws-regulations/summary-toxic-substances-control-act.

[23] Gail A. Van Norman. "Limitations of Animal Studies for Predicting Toxicity in Clinical Trials: Is it Time to Rethink Our Current Approach?" In: *JACC: Basic to Translational Science* 4.7 (2019), pp. 845–854. DOI: 10.1016/j.jacbts.2019.10.008. URL: https://pmc.ncbi.nlm.nih.gov/articles/PMC6978558/.

Chapter 4

Paints

Imagine you are seated in an average living room in an average American home. In all likelihood, all of the walls and ceilings are painted. Among all materials, paints have the most surface area with which to directly interface with the indoor air. If a toxin lives in a paint, that toxin has an exceptionally large platform from which to jump off. Thus, the subject of proper paint selection and application is the first (and maybe most important) chapter in this book among those that deal with particular materials.

I would like to extend a special thanks to Sam Goldberg for providing valuable insights about the paint industry in support of this chapter.

Sam Goldberg is the president of American Formulating and Manufacturing (AFM) and one of the founders and managing members of VitiSeal International, LLC. Mr. Goldberg received a BA with highest distinction from Northwestern University and graduated cum laude from Harvard Law School,. After several years practicing business law, the opportunity arose to manage AFM's transition from a specialized company making dedicated products for environmental medicine physicians and their patients, chemically sensitive individuals, to one using sophisticated technology to make safer products for everyone. Because of his support for environmental issues and belief that effective change can best be accomplished within existing structures and the marketplace, Mr.

Goldberg has worked to make AFM into a company that helps solve indoor air pollution and chemical toxicity problems without requiring sacrifices from, or changing the behavior of, consumers. For over 35 years AFM's mission has been to provide building and maintenance products that are formulated to be as safe and not toxic as possible from a human health standpoint. It was selected by Inc Magazine as one of the top 50 "green" companies in the country, and its Safecoat® products have a strong reputation as one of the patriarch brands in this growing market. AFM has received a Special Commendation for Environmental Achievement and an Earth Award from the City of San Diego, where it is headquartered, and the AMA's Edison Award for Best Environmental Products.

4.1 Executive summary

For any new construction, I recommend pure mineral paint for its safe ingredients and high permeability.

For any existing builds and/or finished wood, I recommend applying AFM Safecoat over surfaces with existing latex paint. AFM Safecoat is suitable for this use due to its high compatibility with other latex paints, low toxicity profile, and low permeability. Existing latex paint can be hazardous due to improper application or the intrinsic toxicity of its ingredients.

I strongly advise against oil-based paints for interior usage. I will not discuss oil-based paints in this chapter because modern oil-based products are almost never a viable option for interior walls and ceilings.

For homes built before 1978, which may contain lead-based paint, I recommend professional testing and strict compliance with all applicable federal, state, and local regulations. I do not provide specific guidance on lead remediation or encapsulation due to the complexity and variability of laws governing these activities.

4.2 Types of paint

First, let me declutter some jargon. The only paints I will recommend purchasing in this chapter are latex paints and mineral

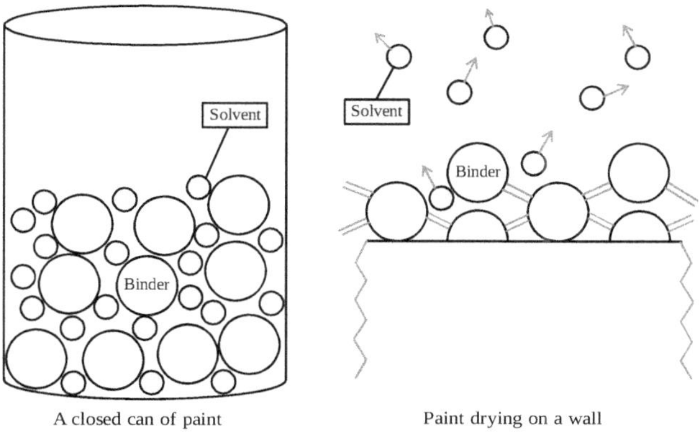

Figure 4.1: Paints and solvents

paints. The term *latex* is a generic name for a synthetic polymer called *acrylic*. Acrylic is a stiff and transparent plastic. Latex paint has no substantial relationship to natural latex. Hereafter, I will use the term *latex* to refer to synthetic latex used in commonly available paints. This also would include the new-to-market *bio based* paints, as these still contain synthetics.

The etymology of latex is illustrative. The word *latex* is merely a stylization of the word *lattice*, in reference to the lattice-like chemical structure of the polymer chains in latex products. In this sense, latex is molecularly indistinguishable from plastic. Similarly, we can say that *latex* and *plastic* are both generic names for synthetic polymers. Synthetic polymers are the distinguishing component of latex paints.

Potassium silicate ("water glass") is the distinguishing component of mineral paint. Water glass is formed by reacting sand (a.k.a. quartz or silicon dioxide) with potash (i.e. natural potassium deposits) at temperatures around 1800 degrees Fahrenheit. The result is a soup of liquid that is capable of binding to mineral surfaces in a microcrystalline structure. The resulting paint is very long-lasting, breathable, and naturally antimicrobial. Mineral paint is long-lasting because it resists discoloration from ultraviolet light. It is breathable because it forms a rock-like porous surface on the wall. And, it is antimicrobial because it maintains a pH of around 11.5 before and after it cures.

For clarity, mineral paint made of water glass is not limewash or lime paint. I will address limewash separately when discussing treatment of stud wall cavities.

4.3 How paint works

Paint has three main components. Every paint needs a binder, a pigment, and a solvent. Everything else is an additive. The binder is the structural substrate of the paint. For example, in latex paints, the binder is latex. In mineral paints, the binder is potassium silicate. The pigment gives the paint its color. For example, titanium dioxide is an abundantly common white pigment used in all sorts of paints. (Even non-white paints use titanium dioxide as a sort of baseline color to which other non-white pigments are added.)

The solvent is responsible for keeping the paint liquid in the can while still allowing the paint to cure into a solid film on the wall. Broadly, a solvent is a chemical that facilitates the dissolution of some other chemical in liquid. Since paints must be wet in the can but dry on the wall, solvents used in paints must be *volatile*. The volatility of a solvent refers to its tendency to evaporate into the air easily. In chemistry, we might characterize a volatile compound as one with a low boiling point (i.e. it evaporates at moderate temperatures) and a high vapor pressure (i.e. it evaporates eagerly). If the solvents used in paints were not volatile, the paint would not dry on the wall, or the paint might require some additional outside force to dry on the wall. Simple solvents that can be used in paints include propylene glycol, acetone, and water.

For a paint to serve its decorative purpose, the binder must entrap the pigment during the curing process. Further, the binder must create a *film* on the wall. Generally, a good looking film is (1) thin, (2) continuous, (3) uniformly textured, and (4) solid. Only high-quality films properly entrap the pigment. Low-quality films allow the pigment to slough off in a fine dust when agitated. Films necessarily trap additives as well. Many of these additives

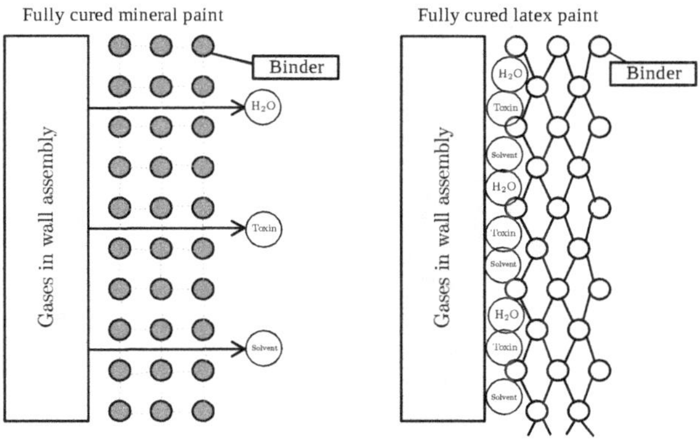

Figure 4.2: The permeability of mineral and latex paints

have beneficial mechanical properties. Some additives are benign. Others are toxic.

A properly cured paint will leave behind a film without solvents. A film with leftover or entrapped solvents has been improperly cured. An improperly cured film does not (1) release all of its solvents or (2) effectively entrap its additives. Improperly cured films are the greatest paint-related hazards to homeowners.

4.4 Improperly applied paint

The inertness of cured paint requires perfect application. Theoretically, a perfectly applied paint releases all of its solvents to form a hard film on the wall that completely entraps all pigments and additives. In reality, perfect application is very rare.

The average paint requires two coats, possibly atop a separate primer. Each coat needs a specific amount of time to fully cure under specific environmental conditions. Requirements will vary according to formulation. For example, a latex paint from 50 years ago with highly volatile solvents will likely require less time to fully cure than a modern zero-VOC latex paint. Additionally, the older paint will be more tolerant to variations in relative humidity and ambient air temperature. Almost all paints require significant ventilation during the curing phase to encourage solvent release. In my experience, painters are typically ignorant of these varying requirements. They will typically apply the next coat as soon as the previous coat can handle it (necessarily before a complete cure) Prematurely applied coats block the curing process of the underlying coats. Partially cured coats fail to completely release their solvents. Accordingly, partially cured coats form an imperfect film. Imperfect films fail to effectively entrap pigments and additives, creating the conditions for off-gassing.

4.5 Off-gassing

Paint interfaces with indoor air. In colloidal science, this means that paint releases chemicals into the indoor air, and the indoor air releases chemicals into the paint.[3] When paint cures, it releases solvents into the indoor air. This is an intended interface. As discussed earlier, paint generally has an enormous surface area within a home. This makes it an important material of concern for human health. Accordingly, I am very interested in understanding the potential for unintended interfaces between the indoor air and paint.

The term *off-gassing* describes an unintended interface between paint and the indoor air. Off-gassing describes the release of chemicals from the paint into the indoor air after partially or fully curing. For clarity, off-gassing does not refer to the evaporation of solvents into the air during the curing process.

Partially cured paint off-gasses significantly more than fully cured paint.[5] Further, partially cured paint off-gasses chemicals with which humans were never intended to cohabitate. In perfect conditions, fully cured paint entraps all ingredients except the solvents. In partially cured paint, homeowners must contend with the possibility that their indoor air is polluted by every chemical in the paint, including the binders, pigments, solvents, and additives. In particular, homeowners should be concerned about the off-gassing of (1) toxic solvents, (2) bioactive additives (e.g. fungicides), and (3) unknown decomposition products. Off-gassing can be accelerated by high temperatures, high humidity, ultraviolet

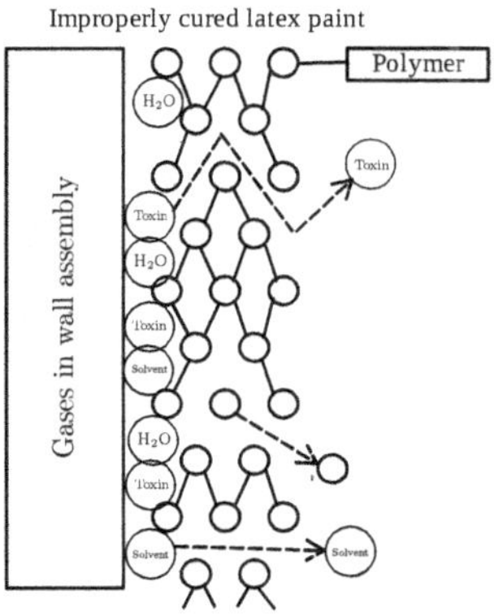

Figure 4.3: Conceptual model of off-gassing potential from improperly cured latex paint

light, simple spills, and physical agitation.

4.6 The biocide additive

When I talk about off-gassing and its health impacts, I am interested in avoiding exposure to an unknown, complex, and synthetic chemical soup. Still, there is one class of chemical additives that is particularly concerning. Virtually all latex paints contain biocides. Biocides are specifically designed to be bioactive. Biocides are a broad class of compounds specifically designed to kill or impair biological life, including pesticides, herbicides, and fungicides. All biocides, even those targeting specific organisms, should be assumed to be harmful to human health in the absence of strong evidence to the contrary. History is replete with examples of humans suffering adverse effects from exposure to biocides that target non-human organisms.

Virtually all latex paint contains some form of fungicidal or bacteriocidal additive. If paints did not contain these additives, they would spoil when stored at room temperature, just like fresh food. For some reason, latex paints are almost universally feedstock for mold and bacteria. They can be canned in a sterile environment, just like canned food, but they will spoil in storage if microbes are introduced to the can during use. If a latex paint does not require refrigeration in the can, then should be assumed to contain biocides.

Some paints also contain fungicides designed to resist mold after curing. For example, a paint used in a full bathroom should expect regular exposure to high humidity levels. These paints commonly contain fungicides designed to resist mold formation during the use phase.

As homeowners, there are a lot of chemicals to be concerned about from off-gassing from partially cured paint. Biocides are ubiquitous in latex paint. In a world of incomplete information, we only need enough information to make a decision. The ubiquity and specific hazard presented by biocides in latex paints is an effective lever against which to evaluate our options. Accordingly, any latex paint we recommend must have a reasonably safe biocide specifically engineered for the sole purpose of preserving freshness in the can.

4.7 VOCs in paint marketing

As of the time of writing, the marketing of supposedly healthy and environmentally friendly latex paints is dominated by language based on VOC content. The acronym *VOC* stands for volatile organic compound. VOC-based marketing is a false premise. Paint companies that use marketing claims about VOC content are exploiting a common misunderstanding about the relationship between VOC content and safety. For a variety of reasons we will discuss, homeowners often falsely associate low VOC content with safety. Marketing claims about VOC content should be completely disregarded due to a lack of definiteness as to their meaning.

First, the Federal Trade Commission has successfully prosecuted claims that paint companies played too fast and loose with their VOC-based marketing. Namely, paint companies have made claims about VOC content in tinted paints that are only valid for base paints.[2] Additionally, paint companies have made untested VOC-based claims about VOC contents during both the application and use phase.[1] In the same advertising, at least one company made implicit claims as to the relationship between VOC content and safety, while implying particular safety benefits to pregnant women and children.[1] These cases and their associated advertisements suggest that paint companies are keenly aware of this false premise and seek to exploit it for marketing purposes.

Second, the definition of a VOC is unclear when used in paint marketing. In chemistry, a VOC refers to a volatile organic compound. Each of those three words means something very specific. The volatility of a compound refers to its tendency to evaporate eagerly at low or moderate temperatures. In chemistry jargon, we might say that a volatile substance has a high vapor pressure (i.e. evaporates eagerly) and a low boiling point (i.e. evaporates at low or moderate temperatures). The organicness of a compound refers to the presence of carbon atoms in the compound. In other words, the *O* in *VOC* merely suggests that the chemical compound has a carbon atom and that one might study it in organic chemistry class. The word *compound* is merely a synonym for a chemical or a molecule.

A chemical's classification as a VOC has no relationship to its inhalation toxicity, which is all I care about when discussing the safety of paints. For example, the VOC in vodka (ethyl alcohol) is generally safe to inhale, while the VOCs in gasoline (benzene, toluene, ethylbenzene, and xylene) are extremely toxic. Organic compounds are distinguished from inorganic compounds. Common inorganic compounds include many acids, bases, salts, and metals. Common organic compounds include nutrients, alcohols, and fuels. This concludes my discussion of the purely chemistry-based definition of a VOC.

When we read marketing statements related to VOCs, it is unclear whether or not we are reading a statement based on the chemistry definition or the regulatory definition. A prominent regulatory definition of VOCs comes from the Clean Air Act. The regulatory definition is oriented towards the reduction of ground-level ozone (i.e. smog) to protect ambient air quality. A VOC that does not react to ultraviolet light will necessarily not contribute to smog. This is why the Clean Air Act specifically exempts non-photochemically reactive compounds from VOC regulation.[6] These non-smog-producing compounds are called *exempt compounds*.[8,13,11] Paint companies are required to disclose the regulatory VOC content of paints in grams per liter on product labels. Accordingly, the regulatory VOC content of a paint only captures the content of non-exempt VOCs. A paint company that can leverage the regulatory definition of a VOC to achieve a regulatory VOC content of zero grams per liter will likely also market that product as "zero VOC".

The smog-producing ability of a specific VOC has no direct relationship with the safety of that VOC as an ingredient in a building material. Propylene glycol (PG) is an example of a non-toxic solvent that is both a chemical VOC and a regulatory VOC. (Roll-on deodorants can be more than 50% PG by weight.[4]) Therefore, we as homeowners should disregard claims about the low or nil VOC content of paints on the basis that they are not directly related to safety.

Theoretically, under this regulatory regime, an extremely dangerous chemical VOC can be used in a product with a regulatory VOC content of zero grams per liter. Methylene chloride is a real-world example of this type of chemical. In 2024, the EPA leveraged its powers under Toxic Substances Control Act (the TSCA) to effectively ban methylene chloride.[9,12] Methylene chloride was primarily used in paint strippers and caused dozens of deaths from inhalation during application. It is a great irony that the EPA had to use the TSCA rather than the Clean Air Act to ban a harmful VOC. It is also ironic that methylene chloride is exempted by name in the black-letter law of the Clean Air Act (see Title 40, section 51.100(s)(1) of the Code of Federal Regulations). (This suggests that Congress was pressured by lobbyists to specifically exempt methylene chloride from the Clean Air Act rather than leave the question up to the EPA.)

Practically, paint companies are required to report VOC contents sans exempt compounds using a mathematical formula specified in Title 40, section 59.406 of the Code of Federal Regulations.[7] As of the time of writing, some paint brands are still using this calculation as a justification to market their products as "Zero VOC". In those same products, VOC disclosures on the labels do not indicate zero grams per liter of VOCs. Rather, they tend to indicate "VOC: <50 g/L as per 40 CFR 59.406". This indicates a reluctance to disclose the exact VOC content on the back of the paint can despite marketing "Zero VOC" on the front of the paint can. This is concerning because there exists semi-hazardous exempt VOCs that are known to be used in paint, including acetone and t-butyl acetate.[10] Additionally, there exists semi-hazardous solvents commonly used in paints that are not definitionally VOCs, including ammonium carbonate (a precursor to ammonia).

For clarity, I am not suggesting here that any paint company used this particular marketing tactic with methylene chloride. Rather, I am suggesting that too many loopholes exist by which paint manufacturers can include hazardous ingredients in products that leverage low-VOC marketing. Accordingly, we should have no confidence in low-VOC marketing.

Thirdly, regulatory VOC-based marketing defies common sense and the fundamentals of proper paint application. The regulatory definition of VOCs combined with consumer demand for low-VOC paints discourages paint manufacturers from using safe and effective VOCs in paints. In the world of paints, VOCs are solvents. We know that all paints need solvents so that they are wet in the can and dry on the wall. We know that proper curing requires complete solvent evaporation. We know that rapid solvent evaporation requires good ventilation. We know that a properly cured paint contains no leftover solvents. And, finally, we want to achieve a complete cure and a tough film in order to entrap all known and unknown additives that may interface with indoor air and ultimately harm our health.

Accordingly, the ideal solvent is both highly volatile and safe to apply under moderate ventilation. That ideal solvent is propylene glycol (PG). Unfortunately, the paint industry is in a race to minimize regulatory VOC content in order to appeal to health-conscious (albeit slightly misguided) consumers. In doing so, it is likely to gravitate to one of three solvent options: (1) exempt VOCs, (2) low-volatility synthetics, or (3) water. The current regulatory and business environment pressures paint manufacturers to avoid logically optimal solvents like PG.

4.8 The Decision Process

In consideration of all of the forgoing, and acknowledging my limited access to information about all available products, I make the following recommendations. For any new construction, I recommend pure mineral paint for its safe ingredients and high permeability. For any existing builds, I recommend applying AFM Safecoat over surfaces with existing latex paint.

Pure mineral paints use inorganic pigments in silicate binders with water as a solvent. The cured film is a solid microcrystalline structure. Pure mineral paints contain no added biocidal ingredients, but they have natural and safe antimicrobial properties deriving from the high pH of potassium silicate. Pure mineral paints have high breathability. This breathability distinguishes them from latex paints.

Breathable paints are preferred in new constructions to discourage mold growth. Non-breathable paints are only preferred when there exists a need to trap toxins into the wall cavity. I will discuss breathability and mold prevention in a later chapter on wall systems.

AFM Safecoat is a brand of latex paint. I recommend AFM Safecoat to my clients as the safest latex paint known to me in cases where (1) compatibility with an underlying latex paint is required and (2) there exists a theoretical benefit to entrapping toxins from the underlying latex paint. AFM Safecoat has the following properties that make it a standout among latex paint brands.

All AFM Safecoat products use water as the sole solvent. (Consequently, they are marketed as zero-VOC.) The binders used by AFM Safecoat are of a high quality, ensuring a tough film given proper application. The tough film is effective at entrapping toxins that may be present in the paint as well as encapsulating toxins beneath the paint, which may include other paints or synthetic building materials. AFM Safecoat paints use a mild biocide designed solely to keep the paint fresh in the can at room temperature. The biocides in AFM Safecoat paints are not designed to be active during the use phase. AFM Safecoat does not explicitly disclose its ingredients. Despite this lack of transparency, I feel confident in recommending AFM Safecoat because it has a long track record of success with my chemically sensitive clients.

References

[1] Federal Trade Commission. *Paint Companies Settle FTC Charges That They Misled Consumers; Claimed Products Are Emission- and VOC-free and Safe for Babies and other Sensitive Populations.* Press Release. Released: July 11, 2017. Accessed: 2025-12-23. 2017. URL: https://www.ftc.gov/news-events/news/press-releases/2017/07/paint-companies-settle-ftc-charges-they-misled-consumers-claimed-products-are-emission-voc-free-safe.

[2] Federal Trade Commission. *Sherwin-Williams and PPG Settle FTC Charges That They Misled Consumers to Believe Their Paints Were Free of Potentially Harmful Volatile Organic Compounds.* Press Release. Released: October 25, 2012. Accessed: 2025-12-23. 2012. URL: https://www.ftc.gov/news-events/news/press-releases/2012/10/sherwin-williams-ppg-settle-ftc-charges-they-misled-consumers-believe-their-paints-were-free.

[3] Georgios M. Kontogeorgis and Kiil Søren. *Introduction to Applied Colloid and Surface Chemistry.* Wiley, 2016.

[4] National Center for Biotechnology Information. *Propylene Glycol - Toxicity.* PubChem Compound CID 1030. PubChem database maintained by NCBI. Accessed: 2025-12-23. U.S. National Library of Medicine, 2025. URL: https://pubchem.ncbi.nlm.nih.gov/compound/1030#section=Toxicity.

[5] Alexandra Schieweck and Marie-Christin Bock. "Emissions from low-VOC and zero-VOC paints – Valuable alternatives to conventional formulations also for use in sensitive environments?" In: *Building and Environment* 85 (Feb. 2015), pp. 243–252. ISSN: 0360-1323. DOI: 10.1016/j.buildenv.2014.12.001.

[6] South Coast Air Quality Management District. *Exempt Compounds.* Accessed: 2025-12-23. South Coast Air Quality Management District, 2024. URL: https://www.aqmd.gov/home/rules-compliance/compliance/vocs/exempts.

[7] U.S. Environmental Protection Agency. *40 CFR 59.406 – Compliance Provisions.* Tech. rep. National Volatile Organic Compound Emission Standards for Architectural Coatings. Current as of December 2025. U.S. Government Publishing Office, 2025. URL: https://www.ecfr.gov/current/title-40/chapter-I/subchapter-C/part-59/subpart-D/section-59.406.

[8] U.S. Environmental Protection Agency. *40 CFR § 51.100 – Definitions.* Current as of December 2025. Accessed: 2025-12-23. 2025. URL: https://www.ecfr.gov/current/title-40/chapter-I/subchapter-C/part-51/subpart-F/section-51.100.

[9] U.S. Environmental Protection Agency. *Biden-Harris Administration Finalizes Ban on Most Uses of Methylene Chloride, Protecting Workers and Communities from Fatal Exposure.* News Release. Released: April 30, 2024. Accessed: 2025-12-23. 2024. URL: https://www.epa.gov/newsreleases/biden-harris-administration-finalizes-ban-most-uses-methylene-chloride-protecting.

[10] U.S. Environmental Protection Agency. *Complete List of VOC Exemption Rules.* Last updated: June 25, 2025. Accessed: 2025-12-23. 2025. URL: https://www.epa.gov/ground-level-ozone-pollution/complete-list-voc-exemption-rules.

[11] U.S. Environmental Protection Agency. *Revision to the Regulatory Definition of Volatile Organic Compounds.* Accessed: 2025-12-23. 2025. URL: https://www.epa.gov/ground-level-ozone-pollution/revision-regulatory-definition-volatile-organic-compounds.

[12] U.S. Environmental Protection Agency. *Risk Management for Methylene Chloride.* Last updated: November 13, 2025. Accessed: 2025-12-23. 2025. URL: https://www.epa.gov/assessing-and-managing-chemicals-under-tsca/risk-management-methylene-chloride.

[13] U.S. Environmental Protection Agency. *Volatile Organic Compound Exemptions.* Last updated: June 13, 2025. Accessed: 2025-12-23. 2025. URL: https://www.epa.gov/ground-level-ozone-pollution/volatile-organic-compound-exemptions.

Chapter 5

Engineered Wood

Throughout this book, I have made many recommendations that imply the usage of engineered wood products. Engineered wood is distinguished from solid wood by the use of chemical binders and additives. Most engineered wood products a homeowner will commonly encounter during construction use well-understood chemicals. Engineered woods have useful structural properties I will discuss.

Engineered woods are inevitable in modern construction and interior design. Most of them use chemicals of concern like formaldehyde and isocyanates. Even if we wished to avoid them, we could not. Thankfully, there exist many ways to responsibly and safely use many engineered woods. I will discuss these at length. Homeowners are thus confronted with the difficult challenge of developing literacy in engineered wood products.

5.1 Definitions

In order to deliver clear guidance on how to use these products, I need to clearly define a few classes of products according to their use cases. There exist four main use cases of engineered wood products: (1) structural paneling, (2) general interior, (3) bathroom, and (4) kitchen. Structural paneling includes almost all sheet-like wood that is invisible to the occupant, including sheathing and subfloors. General interior usage is all interior usage that is not in a kitchen or bathroom. General interior usage is distinguished from kitchen and bathroom usage by its relatively unchanging environmental conditions. Bathroom usage is expected to undergo dramatic swings in relative humidity levels, often sustaining high relative humidity levels for long and regular periods of time. Finally, kitchen usage is expected to undergo dramatic swings in both temperature and relative humidity during regular usage, often sustaining high temperature and high relative humidity simultaneously. The three non-structural classifications mainly serve to describe cabinets, vanities, moldings, furniture, and finished flooring.

Next, I will need to define a few classes of engineered wood products. First, Urea-formaldehyde ("UF"), melamine-urea-formaldehyde ("MUF"), and melamine-formaldehyde ("MF") are classified as low-stability resins. Then, I will define phenol-formaldehyde ("PF") and methylene diphenyl diisocyanate ("pMDI") as high-stability resins. High-stability resins are distinguished from low-stability resins by their chemical stability under high heat and high relative humidity. I define bio-based resins to include any resin that does not rely on petrochemicals. I will refrain from classifying bio-based resins as low or high stability based on a lack of standardization and data. Some bio-based resins achieve good emissions performance and moisture resistance, but others do not.

Now, I need to define some broad classes of engineered woods according to their total chemical binder content. Very generally, all engineered woods are made of small pieces of wood combined with resin. As these pieces get smaller, the required resin content gets higher. As the total resin content gets higher, the health risks posed by resin off-gassing increase. I define three classes of resin content: (1) solid wood, (2) low-resin and (3) high-resin. Solid woods are not engineered and have no resin. Low-resin products use relatively large wood pieces so require less resins than others. I will define low-resin to mean typically below 6% total resin content. Important low-resin products include plywood and oriented strand board ("OSB"). Both typically contain 3% to 5% resin content by weight for structural products. Plywood is a stack of solid veneer sheets glued together. OSB is made of strategically laid strands of wood that are about 4 inches long and a little over 1 inch wide, combined with resins. Other products like laminated veneer lumber ("LVL"), parallel-strand lumber ("PSL"), and cross-laminated timber ("CLT") can also meet the definition of low-resin. I define high-resin to mean any wood product that is generally 8% to 15% resin content. Important members of this class include medium-density fiberboard ("MDF"), high-density fiberboard ("HDF"), particleboard.[3] These products are made of relatively small fibers and particles. Accordingly, they require more binders to hold them together.

Finally, I define *high-quality* engineered woods to mean any engineered wood that is both high-stability and low-resin. Exterior-grade and structural-grade plywoods and OSBs will almost always fall into this category. Extra verification is encouraged for all other types of engineered wood.

I define *non-formaldehyde* to refer to resins without formaldehyde in the chemical formulae of the principle components of the resin. Per my definition, non-formaldehyde wood products use either pMDI, bio-resins, or solid wood. I define *sealed* to mean any wood product that is fully sealed through some combination of acrylic and veneering.

5.2 Executive summary

For structural paneling in new construction, I recommend high-quality engineered wood. I advise against using solid wood in place of structural paneling generally. I strongly recommend against inhabiting houses with low-stability structural paneling.

For interior general use, I recommend (1) solid wood, (2) high-quality veneer-based products (namely plywood), and (3) trustworthy bio-based products. I advise against low-stability products for general interior use. I recommend incrementally replacing low-stability products in interior general spaces.

For bathroom use, I recommend sealed solid wood and sealed non-formaldehyde products. For kitchen use, I recommend sealed solid wood and sealed pMDI products. If required to use non-pMDI resins in a kitchen, I recommend taking extra precautions to protect them from heat and humidity. I recommend completely replacing low-stability products in existing bathrooms and kitchens.

In all applications, I recommend against new purchases of low-stability engineered wood products. I recommend maintaining proper indoor air temperature, relative humidity, and ventilation to mitigate the potential health impacts of off-gassing from engineered wood products. (See the chapter on HVAC design for more information.)

5.3 Low-stability products

Historically, concerns regarding the toxicity of engineered wood products have focused on gaseous emissions of formaldehyde. In Chapter 3, I discussed the historical context in which the EPA failed to protect the public from formaldehyde emissions over the past 50 years. As of the time of writing (early 2026), all engineered wood products sold in the U.S. must comply with TSCA Title VI or CARB ATCM Phase II for formaldehyde emissions. To understand what this means for us as homeowners, I need to discuss the nuts and bolts of these laws and regulations. Then, I need to discuss how the industry has responded and provided homeowners with alternative options.

According to EPA TSCA Title VI (40 CFR Part 770), all composite wood products sold in the U.S. (including laminated products after the exemption expired in 2024) must comply with TSCA Title VI. TSCA Title VI is harmonized with CARB ATCM Phase II, so I will focus on TSCA Title VI for simplicity. TSCA Title VI sets different maximum formaldehyde emission requirements according to the type of engineered wood per ASTM E1333-14 (Table 5.1).

Product	**Limit**
Plywood	0.05 ppm
Particleboard	0.09 ppm
Non-exempt OSB	0.09 ppm
MDF	0.11 ppm
Thin MDF	0.13 ppm

Table 5.1: Formaldehyde emission limits

ASTM E1333-14 is a testing method for measuring formaldehyde emissions. The test is carried out as follows. A representative sample of the engineered wood product is placed in a large environmentally controlled chamber. The size of the sample is cut such that it is 0.95 square meters per cubic meter of chamber space. Purified air is pumped into the chamber at a rate of 0.5 air changes per hour ("ACH"). Within the chamber, the temperature is held at 77°F and the relative humidity is held at 50%. Formaldehyde concentrations within the chamber are continuously monitored until a steady state is reached. The 95% upper confidence limit estimate of the formaldehyde concentration must be below the associated threshold for the product to pass the test.[2]

Formaldehyde can be off-gassed in a few ways. First, unreacted formaldehyde from urea-formaldehyde binders can simply evaporate off of the material. Second, a process called hydrolysis allows cured urea-formaldehyde resin to react with water in the air to decompose back into formaldehyde and urea. This process causes gaseous formaldehyde to be released into the air.[4] A simplified representation of the hydrolysis reaction is reproduced from Frihart (2020) below. (See citation at end of chapter.)

$$\text{Formaldehyde} + \text{Urea} \xrightleftharpoons{\Delta\ +\ H^+} \text{Urea-Formaldehyde} + \text{Water}$$

Figure 5.1: "The reversible reaction of urea and formaldehyde for providing the urea-formaldehyde polymer that can be hydrolyzed to liberate formaldehyde."[4]

The logic of the ASTM E1333-14 test takes these two chemical mechanisms into account. The ASTM test is performed a week after the engineered wood is produced in order to allow the bulk of the unreacted formaldehyde to off-gas. This is a realistic parameter of the test, because most engineered wood products will sit for weeks to years in factories, boats, warehouses, and retail shelves before sale.

Therefore, ASTM E1333-14 is primarily testing for the quantity of formaldehyde that is released by hydrolysis at a particular combination of temperature, humidity, and ventilation. Manufacturers have a number of levers available to control formaldehyde emissions in their products. For example, they can (1) add acrylic sealants, (2) add formaldehyde scavengers to the resin, (3) cure the resin at higher temperatures, or (4) use higher-quality resins. These are merely a few examples. The engineering options available to engineered wood manufacturers are endless. Regardless of the ingredients and processes used, the product still has to pass this test (if not separately qualified for a testing exemption).

As reasonable homeowners, we are grateful for the advances in these sorts of regulations, but there are gaps in the testing we need to be aware of. The following facts are pertinent. While the reservoir of unreacted formaldehyde in the product is finite, the reservoir of unhydrolyzed formaldehyde-based resin in the product is virtually inexhaustible over its lifespan. Consider the following. The unreacted formaldehyde is highly volatile. In chemical terms, it is a VOC. It has a low boiling point and a high vapor pressure. The vast majority of unreacted liquid formaldehyde left in the pores or crevices of engineered wood will evaporate away in a few weeks after its production. The unreacted liquid formaldehyde is likely a very small fraction of the total weight of the product. Some estimates say the formaldehyde content of engineered woods is at around 0.1% by weight, so the unreacted formaldehyde would naturally be a minuscule fraction of that. Then, the unreacted liquid formaldehyde present after the product reaches its destination would be yet another minuscule fraction of that. Accordingly, it is not of significant concern to us.

The potential emissions of unhydrolyzed formaldehyde are unending by comparison. ASTM E1333-14 tests mandate a specific exchange rate of air under specific environmental conditions. ASTM E1333-14 sets these parameters at reasonably normal levels. Indoor conditions of 0.5 air changes per hour at 50% relative humidity and 77 degrees Fahrenheit sound like those of a fairly average house. We should still be concerned for two reasons.

First, high-resin products are allowed to off-gas more than low-resin products under TSCA Title VI despite high-resin products having generally lower stability. If plywood is limited to 0.05 ppm but thin MDF is allowed 0.13 ppm, that merely gives us two reasons to worry about thin MDF. Thin MDF will off-gas more than plywood because thin MDF (1) has a higher total resin content and (2) is typically made from low-stability resins. Ideally, TSCA Title VI would hold thin MDF to a higher standard than plywood for ASTM E1333-14 to account for thin MDF's higher off-gassing risk. But, it does the opposite.

Second, the environmental conditions of the ASTM test are unrealistic for kitchens and bathrooms. The ASTM test is probably a good approximation of the off-gassing potential of MDF used in a living room couch, but it under-estimates the off-gassing potential of MDF used in a damp bathroom or a damp-hot kitchen. I have personally tested kitchens where ovens and stoves can heat particleboard cabinetry up to 130 degrees Fahrenheit during normal use. Further, a pot of boiling water can bring the relative humidity of the cabinets above a stovetop to saturation. I am not suggesting that these kitchens were installed correctly or that they are being used correctly. Still, they exist. Off-gassing from the impacted engineered woods in these kitchens is hazardous to occupant health. Further, the reservoir of formaldehyde against which hydrolysis can act is effectively unending in these situations.

To demonstrate this, let me propose a rough example. Consider that a 1,000-kilogram kitchen cabinet set may contain 1 kilogram of chemical formaldehyde. Each time the kitchen is used, a few kilograms of the cabinetry is heated to well above 100 degrees Fahrenheit. In an hour, a few milligrams of formaldehyde are reacted into the air.[4] This would bring the airborne formaldehyde well over 100 ppb, a concentration considered acutely unsafe by multiple regulatory guidelines. In this rough example, millions of cooking sessions would be required to hydrolyze away all of the formaldehyde in the cabinetry. I say that the reservoir of unhydrolyzed formaldehyde is unending because unhydrolyzed formaldehyde will easily outlive the cabinetry and the house itself. (Real-world testing by regulatory authorities has confirmed that normal usage of kitchen appliances can generate indoor formaldehyde concentrations in excess of 300 ppb.[1])

Kitchen cabinets made from low-stability formaldehyde resins are obviously unsafe. They have high surface area, high total formaldehyde content, and regular exposure to extreme temperature and humidity. Kitchen cabinets are a common principle source of indoor air quality issues.

5.4 High-stability products

High-stability resins are more chemically stable than low-stability resins, especially in extreme environmental conditions. I have already discussed the importance of using high-stability products in kitchens and bathrooms. Here, I will discuss the nature of the differences in stability among engineered wood products.

Publicly available research suggests that formaldehyde emissions from UF products can increase 10 to 100 times above their ASTM E1333-14 baselines under extreme environmental conditions.[5,4] Further, the off-gassing properties of MUF mimic the off-gassing properties of UF in the long run, especially under extreme environmental conditions.[7] Formaldehyde emissions testing attempts to identify the presence of the reverse curing reaction (i.e. the hydrolysis reaction) of formaldehyde-based resins.

Accordingly, formaldehyde emissions testing can tell us a lot about PF resins but not a lot about pMDI resins. This consideration is especially relevant to understanding the general risk of hazardous exposure to non-formaldehyde chemicals in these resins. Formaldehyde emissions testing suggests that PF is very stable under normal conditions and moderately stable under extreme conditions.[7] This suggests that PF presents little hazardous formaldehyde exposure and little hazardous phenol exposure in conventional usage. But, it suggests that we should avoid PF in bathrooms and kitchens where high heat and humidity can accelerate hydrolysis.

Publicly available research suggests that pMDI is remarkably stable in both normal and extreme environmental conditions.[7] The principle risk of pMDI degradation is theoretically the release of isocyanates into the air. The principle risk of using pMDI products is the introduction of unreacted isocyanates into the home that reach inhabitants via inhalation and skin contact. Isocyanates are volatile and toxic. Fortunately, isocyanates are well-researched.[6,8] Isocyanate exposure has a long and well-documented history of causing occupational asthma. Further, unreacted isocyanates in consumer products have been a well-monitored source of human health concern for decades. For example, Krone et. al. discouraged the use of polyurethane foam pillows in 2005 because they were believed to contribute to childhood asthma via exposure to unreacted isocyanates. Isocyanate sensitivity is common among manufacturing and construction industry personnel. Yet, many modern cured pMDI products pose no danger even to highly sensitized construction industry personnel, presumably because no excess isocyanate remains in the product. Many products with high pMDI content do not require masks, gloves, or eye protection to cut and manipulate. Further, gaseous isocyanate inhalation is suspected to be less dangerous than liquid isocyanate skin contact, further reducing the hazards of off-gassing to installers and end users.[6]

In a world where engineered wood is unavoidable, pMDI seems to be the best petro-chemical resin. I encourage the industry to move towards more safe usage of pMDI. Still, I encourage more scrutiny and research into pMDI toxicity. Further, I would like to see emissions testing regulations adopted for pMDI that addresses toxicity risks of its component molecules. For clarity, isocyanates are dangerous, and there exists many classes of commercial and consumer products where isocyanates are unacceptably risky. This is especially true for isocyanates that cure on-site (e.g. in spray foam insulation) and isocyanates that produce polyurethane foam (e.g. in mattresses).

Bio-resins are sometimes a desirable alternative to petro-chemical resins. Bio-resins are new and unstandardized. Some have found success in niche applications like cabinetry. Bio-resins frequently lack desirable structural and water-wicking properties of petro-chemical resins. This property disqualifies bio-resins from a lot of structural use cases. Given the current availability of information and the state of the industry, I recommend well-established bio-resins for non-structural and non-heated use cases. In other words, I recommend them for general interior and bathroom applications. Common sense suggests that no generalized hydrolysis reaction exists for bio-resins, so high humidity should not excessively harm them. Common sense also suggests that bio-resins, like most biologic materials, could potentially degrade under simultaneous high heat and high humidity.

5.5 Construction and moisture

In the executive summary for this chapter, I made especially sound advise against the usage of solid wood for applications that would otherwise rely on engineered wood for structural paneling. In other words, I advise against solid wood subfloors. Similarly, I advise against solid wood for exterior sheathing. The reason is quite practical.

Imagine a typical residential construction project. First, the foundation is poured, then the subfloor for the first floor is laid, then the first floor is framed, then the subfloor of the second floor is laid, then the second floor is framed, and so on. The second or third floor will support a roof, which will be framed, then roofed. From the time the first subfloor is laid to the time the roof is completed, the house is wide open to the elements. Namely, it will be rained on. Engineered wood paneling is specifically designed and tested for its moisture wicking abilities. On a palette of engineered wood paneling, you will likely a see a label reading "EXPOSURE 1" or "EXTERIOR".[3] These are designations of resiliency to water penetration. Specifically, "EXPOSURE 1" refers to resiliency to acute temporary water exposure and is suitable for subfloors. "EXTERIOR" refers to resiliency to continuous and repeated water exposure and is suitable for exterior sheathing.

Moisture-wicking properties are important because they discourage mold growth. All wood carries some ambient quantity of mold spores. Exposed wood will acquire more mold spores from the ambient air. Moist wood will grow mold. Worse yet, moist wood that has not had proper time to dry before encapsulation will grow a lot of mold. I believe the health risks posed by potential mold growth in solid wood are greater than the health risks posed by chemical off-gassing from high-quality engineered wood. Accordingly, I recommend against substituting engineered wood structural paneling for solid wood alternatives.

5.6 Identifying engineered woods

As of the time of writing (early 2026), homeowners cannot easily identify the resins in engineered wood products based on common labels. I will offer a few useful tricks here. Ultimately, it may be incumbent on the reader to inquire with wholesalers and manufacturers to address specific questions.

Generally, all modern structural paneling is made with high-stability resins. Any panel labeled "EXPOSURE 1" or "EXTERIOR" is almost certainly made with PF and pMDI. Plywood looks like a thin stack of playing cards, and OSB looks like a squished pile of wood chips.

Homeowners should be suspicious of all high-resin engineered wood products. Generally, high resin content is associated with low particle size. Small particles are only economically relevant because they are inexpensive. Low-stability resins also tend to be inexpensive. Accordingly, an engineered wood product that is designed to compete on price will use small particles and low stability resins. This is helpful to us as homeowners, because small particles are easy to identify. HDF and MDF look uncharacteristically uniform and smooth. The particle size in HDF and MDF resembles that of sand. A cross-section of either looks a little like wet beach sand. Particleboard looks like OSB but has a lower chip density in the center. If it has a single layer of veneer on the top and bottom and a low chip density in the middle of its cross-section, it is almost certainly particleboard. Particleboard is very common in cabinetry and office furniture.

Another class of high-resin low-stability products exists that we have yet to discuss. Wood composites are becoming more popular. Wood composites are defined by their extremely high resin content (typically greater than 40%).[3] They often look like MDF or HDF but will have an unusually high and random quantity of non-wood particulate, almost like sprinkles embedded in a cake. The sprinkles can be any sort of ground up or recycled materials that generally just add mass to the substrate. Obviously, these should be avoided.

The abbreviation "NAF" stands for "no added formaldehyde". It appears on labels of many engineered wood products and suggests the absence of UF, MUF, MF, and PF. The NAF label does not exactly align with any of my specific recommendations, but it is useful for disqualifying products in certain use cases.

The abbreviation "NAUF" stands for "no added urea-formaldehyde". This label appears on products that want to advertise that they do not use UF or MUF. Since MF can be just as problematic as UF and MUF in terms of formaldehyde emissions, the NAUF label does not inspire a lot of confidence. Instead of relying on the NAUF label, homeowners should investigate the binder type by other means. For example, the NAUF label on OSB suggests some mix of PF and pMDI. But, the NAUF label on a box of wooden flatpack furniture suggests a heavy reliance on MF (or even pressed cardboard). In any case, the NAUF label does not provide much valuable information to us because UF and MUF are not the only binders I seek to avoid.

The abbreviation "ULEF" stands for "ultra-low emission formaldehyde". It appears on the labels of many engineered wood products. Generally, ULEF is used on MUF, MF, and modified UF products. It can be used on PF products. Generally, a ULEF label suggests the presence of a low-stability binder. Research data suggests innovations like MUF, MF, modified UF, and formaldehyde scavengers do not materially improve emissions profiles in extreme environmental conditions.[7,4] Accordingly, I assume a ULEF label is suggestive of a low-stability binder unless the presence of PF is proven otherwise. In other words, a ULEF label on anything other than structural sheathing indicates a low-stability binder. Still, the ULEF label does not perfectly align with any of my categories of engineered woods, so it should not be overly relied upon.

The biggest purchasing mistakes made by homeowners are often made online and in the aisles of big box stores. These quick tips and tricks will help homeowners disqualify many products without the need to consult with the retailer or manufacturer. But, big purchases deserve transparency, and homeowners should feel empowered to require detailed data about material composition from retailers or manufacturers. Homeowners are encouraged to use the Declare List to check for specific ingredients in specific products.

5.7 Finish carpentry, cabinetry, and furniture

At the beginning of this chapter, I recommended incrementally replacing low-stability products in general interior spaces. I also recommended replacing all low-stability products in kitchens and bathrooms. In the previous chapter, I recommended painting over existing hazardous latex paint with AFM Safecoat. All of this guidance collides in an unpleasant way when we consider finish carpentry. Finish carpentry includes trim, mouldings, and baseboards. Readers may also find it difficult to reconcile this advice with regards to their existing cabinetry and furniture. Worse yet, what about hardwood floors with HDF cores? I will address these edge cases here.

My goal is to eliminate as much chemical off-gassing as possible with reasonably achievable upgrades. In occupied homes, only one upgrade can be made at a time. Also, each upgrade takes time. We as homeowners need to be thoughtful about how we sequence these upgrades. To eliminate chemical off-gassing from engineered woods products, I have three tactics: (1) remove, (2) replace, and (3) seal.

When sequencing upgrades, I consider the following key facts. (1) The off-gassing potential of a product is proportional to the weight of its total binder content. (2) The off-gassing potential is magnified by high temperature and high humidity. (3) Most paint-like coatings will reduce off-gassing by sealing chemicals inside the material. (4) Generally, removal takes minutes, sealing takes hours, replacement takes weeks.

I will propose a sequence of upgrades for a hypothetical average occupied home in the U.S. (the "model home"). In this model home, no consideration has been made for the safety of engineered woods used throughout the home. Therefore, all materials selections have been made according to what is common. In this model home, I would sequence upgrades related to engineered woods in the following order. (1) Remove all non-essential furniture made of engineered wood. (2) Seal all interior engineered wood surfaces that may come into contact with heat or humidity using the appropriate AFM Safecoat products. [This step targets cabinetry and finish carpentry in kitchens and bathrooms.] (3) Replace all remaining large furniture made of engineered wood. (4) Seal any unsealed interior engineered wood surfaces. This step targets remaining furniture and finish carpentry throughout the home that is not already sealed by polyurethane sealant or latex paint. (5) Replace the cabinetry in the kitchen and bathroom. (6) Follow other recommendations in this book related to carpeting, dehumidification, and ventilation. (7) Test the formaldehyde in the air. (8) Explore further removal of engineered woods if tests indicate high levels of formaldehyde.

Notice how I did not recommend replacing exterior sheathing or subfloors. The OSB in a house is large, ubiquitous, encapsulated, immovable, and relatively stable. It will off-gas formaldehyde over its life, both from the PF resin and from the wood itself. This off-gassing will create an unavoidable baseline exposure that is only addressable by filtration and ventilation. This quantity of formaldehyde off-gassing is suspected to be tolerable to all but the most sensitive occupants.

Notice how I did not explicitly recommend replacing mouldings, trim, baseboards, hardwood floors, or small furniture items. All of these items are likely to contain some of the lowest quality engineered wood in the home (i.e. MDF and HDF). However, these items are either very small or very encapsulated, so their off-gassing potential is limited. A homeowner should only consider replacing them after the usual suspects have been addressed. (If the baseboard is the culprit, the home probably has a moisture problem rather than a formaldehyde problem.) The formaldehyde off-gassing contribution of these small items is suspected to be marginal compared to the baseline level contributed by the exterior sheathing and subfloor. Therefore, the off-gassing from these items can be effectively addressed by filtration and ventilation.

5.8 Pressure Treated Lumber

Pressure treated lumber is a common source of concern among health-conscious homeowners. Pressure treated lumber has been impregnated with copper-based chemicals that allow it to resist rot from wood-eating fungi. It feels just like normal pine lumber, but it has a distinctive green tint to it, in most instances.

An exploration of the history and toxicity of pressure treated lumber would be of little value. Both the new and legacy versions of pressure treated lumber are impregnated with toxic chemicals. The new version is suspected to be less hazardous, but this does not matter. The new version is still not safe enough for interior use or skin contact. Still, building codes leave homeowners with very little autonomy on this issue. Fortunately, the use cases mandated by building codes are not particularly hazardous to homeowners.

My advice is the following. Avoid using pressure treated lumber unless required by building codes. When it is required, encapsulate it to prevent humans and animals from touching it. Pressure treated lumber can be encapsulated easily in most structural use cases. For other projects, seek alternative materials like metal or rot-resistant hardwoods.

References

[1] ARCADIS Geraghty & Miller, Inc. *Indoor Air Quality: Residential Cooking Exposures.* Tech. rep. Prepared for California Environmental Protection Agency, Air Resources Board. Sacramento, CA: California Air Resources Board. URL: https://ww2.arb.ca.gov/sites/default/files/2020-03/cookingstudy_0.pdf.

[2] ASTM International. *Standard Test Method for Determining Formaldehyde Concentrations in Air and Emission Rates from Wood Products Using a Large Chamber.* Tech. rep. ASTM E1333-14. Reapproved or referenced in 40 CFR § 770.99. West Conshohocken, PA: ASTM International, 2014. DOI: 10.1520/E1333-14. URL: https://www.astm.org/e1333-14.html.

[3] Forest Products Laboratory. *Wood Handbook—Wood as an Engineering Material.* Tech. rep. General Technical Report FPL-GTR-282. Centennial edition. Madison, WI: U.S. Department of Agriculture, Forest Service, Forest Products Laboratory, 2021. URL: https://www.fpl.fs.usda.gov/documnts/fplgtr/fpl_gtr282.pdf.

[4] Charles R. Frihart, Timothy L. Chaffee, and James M. Wescott. "Long-Term Formaldehyde Emission Potential from UF- and NAF-Bonded Particleboards". In: *Polymers* 12.8 (2020), p. 1852. DOI: 10.3390/polym12081852.

[5] Charles R. Frihart et al. "Formaldehyde Emissions from ULEF- and NAF-Bonded Commercial Hardwood Plywood as Influenced by Temperature and Relative Humidity". In: *Proceedings of the International Convention of Society of Wood Science and Technology and United Nations Economic Commission for Europe – Timber Committee.* Paper WS-23. Geneva, Switzerland, Oct. 2010, pp. 1–13. URL: https://research.fs.usda.gov/treesearch/38606.

[6] Cheryl A. Krone and Tom D. Klingner. "Isocyanates, polyurethane and childhood asthma". In: *Pediatric Allergy and Immunology* 16.5 (Aug. 2005), pp. 368–379. ISSN: 1399-3038. DOI: 10.1111/j.1399-3038.2005.00295.x.

[7] Darren Riedlinger, Pierre Martin, and Tom Holloway. *Particleboard Formaldehyde Emissions and Decay under Elevated Temperature and Humidity Conditions.* Tech. rep. March 28, 2012. 475 28111 St., Springfield, OR 97477: Arclin LiSA, Mar. 2012.

[8] U.S. Environmental Protection Agency. *Methylene Diphenyl Diisocyanate (MDI) and Related Compounds Action Plan.* Tech. rep. RIN 2070-ZA15. Accessed: 2025-11-19. Washington, DC: U.S. Environmental Protection Agency, Apr. 2011. URL: https://www.epa.gov/sites/default/files/2015-09/documents/mdi.pdf (visited on 11/19/2025).

Chapter 6

Foundations and Flooring

Imagine you are seated in an average living room in an average U.S. home. Flooring is second only to paint in gross surface area of the materials that surround you. Healthy flooring is just as important as healthy paint when you consider the special characteristics and uses of flooring. Flooring is more likely to be subject to accidental spills that may contribute to chemical degradation or mold formation. Children and pets are physically closer to the floor than adults. Accordingly, they are significantly more exposed to flooring-related toxin inhalation than adults. Flooring can trap dust, dander, mold, pathogens, and chemicals that are brought in from the outside. Some types of flooring (like carpet) trap more inflammogens than others.

6.1 Executive summary

For a foundation in a new construction, I recommend either (1) a basement or (2) a concrete slab on grade. If local conditions and code allows, I recommend a basement for a foundation. If local conditions or code prevent the construction of a basement, I recommend a concrete slab on grade foundation. I advise against crawlspace foundations. While each of these three foundation types are workable if constructed properly, crawlspaces end up being the most problematic. Therefore, I advise against them.

For a new stick frame construction with more than one floor, at least one floor will be an engineered wood subfloor. For an engineered wood subfloor, I recommend a high-quality engineered wood per the definition offered in Chapter 5. Specifically, I recommend high-quality engineered woods over any other alternatives, including solid wood (see Chapter 5 for details).

For existing construction, I recommend removing all synthetic carpet. I recommend replacing all synthetic carpet with any of the flooring recommendations in this chapter. In general, avoiding synthetic carpet is the most important decision for achieving safe flooring.

If using a concrete slab on grade, I recommend finished concrete as the most mold-resistant, cost effective flooring option in general. For any non-concrete subfloor or if you do not want a finished concrete slab, I recommend the following. Porcelain tile is the safest option. Natural linoleum is the next safest option. Then, a floating wood or cork flooring material is the next safest option.

Homeowners that are particularly sensitive to non-toxic odors should avoid cork, linoleum and certain odorous wood flooring. Homeowners should avoid applying wood in bathrooms due to poor moisture resistance.

Homeowners are encouraged to allow their aesthetic preferences to drive the design of flooring decisions in their homes in accordance with this framework. Products recommended here should be selected in a way that minimizes the required total volume of glues, resins, grouts, caulks, and sealants.

For existing homes with crawlspaces, I recommend full encapsulation (a thick polyethylene barrier, minimum 15-mil thick) and climate control (insulation and dehumidification). For existing homes with basements, I recommend fully conditioning any finished or unfinished basement space. Further, I recommend basement interiors be treated with just as much care as other interior spaces in the home.

I recommend that all homeowners test for radon and employ radon mitigation as necessary to achieve radon concentrations below 1 picocurie per liter (pCi/L). I recommend that all homeowners take advantage of powered foundation exhaust systems (like radon fans) to protect themselves from other harmful soil gases as well. Common construction practice is to use a 6-mil vapor barrier under the concrete slab. I recommend a 15-mil vapor barrier below the foundation to directly block soil gas intrusion and to resist punctures during the pour.

6.2 Carpet

Synthetic carpet has a terrible track record on indoor air quality. First, synthetic carpeting is very barely regulated. Numerous toxic chemicals can easily find their way into synthetic carpeting during manufacturing. Most synthetic carpet is a complex mix of polyester and glue. These substances (and their chemical components) often continue to off-gas for a long time, especially when exposed to heat and humidity. History is replete with case studies of newly installed carpeting making occupants sick.[4,1,5,2]

Second, synthetic carpet is a dust sponge. Household dust collects a long list of allergens and toxins. This dust finds its way into the home via interior and exterior sources. Then, it settles on the synthetic carpet. It is then ground into the synthetic carpet and kicked back up into the air through normal use. For example, stomping on a synthetic carpet can release detectable (but often invisible) clouds of dust into the air.[8]

Synthetic carpets are such an extreme source of dust exposure that I recommend homeowners avoid carpets altogether if they do not have access to or cannot afford pure natural carpeting. Pure natural carpeting made from wool, jute, natural rubber and foregoing chemical dyes and pesticides, can be a less problematic alternative to synthetic carpeting. Natural wool fibers, for instance, contain lanolin oil which repels liquids and helps retain household dust until extracted via regular vacuuming with proper equipment. Pure natural carpeting is especially functional for area rugs, runners and areas where a soft feel is appreciated. Think the bedside. However, since carpets are not functionally necessary for a home, I will assume that the full substitution of carpet for other materials is feasible for most homeowners in most areas of the home.

Vacuuming can achieve wildly mixed results in terms of its impact on indoor air quality. Some vacuums kick up carpet dust into the air. Others collect some dust while pulling the rest onto the surface of the carpet, where it is more readily taken up into the air. Wet vacuuming and dry vacuuming barely differ in terms of overall health performance. Wet vacuuming collects some particles more effectively but creates other mold risks in the process. Wet vacuuming also increases chemical exposure through the use of chemical cleaning solutions.

I am not suggesting that those with existing carpet completely avoid vacuuming. Rather, I am emphasizing that there exists no happy way to cohabitate with synthetic carpet. For example, wet vacuuming may be an effective way to address acute issues with pet dander built up in synthetic carpeting. It might even temporarily alleviate allergy symptoms associated with pet dander. Still, it is not a permanent solution.

Synthetic carpeting is the only common flooring option that is mop-like. A synthetic carpet is like a mop that sticks straight up into the air and collects everything in the environment around it. And, it can never really be fully cleaned. Every other flooring option is in some way a flat hard surface. Dust on flat hard surfaces can be effectively removed with a simple wet cloth. Further, water-based cleaning methods applied to hard flat surfaces are not mold risks if applied with the requisite care.

6.3 Glues, Resins, Grouts, Caulks, and Sealants

Flooring products vary wildly in their composition and application. In general, a flooring product can be toxic for all the same reasons that a paint or an engineered wood can be toxic (see earlier chapters on each). For example, certain flooring system might have a continuous glue-based application. These glues should be avoided because of their VOC content, unknown additive content, and unknown long-term off-gassing potential. A product-by-product comparison is unnecessary because of the abundance of healthy alternatives on the market.

Wood flooring systems may contain engineered cores made of low-quality engineered wood. Some may require glue-based installation methods, others use an interlocking system not requiring glue or nails. Obviously, I recommend solutions that do not rely on glues and resins.

Regarding tile, pure ceramic and porcelain are preferred. The only health risks posed by tile in the home is the potential off-gassing of mortar, grout, caulks and sealants, which present the same health challenges as toxic paints. Fortunately, they do not have a large surface area (like painted walls). I also recommend using the largest possible tile to minimize grout, which reduces the area for potential mold growth and off gassing. The exposed grout can be sealed with AFM Safecoat Grout Sealer to control most off-gassing potential. There is also a potential health risk with tile that is cut inside the home, as this can release respirable crystalline silica. Therefore, I recommend that all tile cutting is done outside, away from open windows and doors.

Interior-grade caulks have a lot of general utility in trimming, edge work, and finishings, especially when working on mouldings and flooring. AFM Safecoat caulk is recommended for these use cases. Silicone caulk should be used around wet areas such as tubs or sinks. Silicone presents some off gassing challenge, but the tradeoff is better water and mold resistance.

6.4 Foundations

My top recommendation for new construction is a properly installed and maintained basement. A proper basement is an extension of the interior of the house. Just like the rest of the house, it is waterproof, insulated, and conditioned.

A properly built basement is an achievable and worthwhile investment. The only downside of a basement is the increased risk of moisture intrusion if improperly installed. Basements sit below grade (i.e. below the ground). As such, given an improperly waterproofed foundation, a basement will flood faster than a crawlspace. Once waterproofing is properly addressed, basements offer clear advantages over other types of foundations in terms of indoor air quality.

One of the greatest risks to occupant health is poor execution by trades. In this context, *trades* refers to HVAC technicians, plumbers, and electricians. Basements give trades ample room to work. This decreases the odds of mistakes that can harm occupant health. For example, damaged ductwork can suck toxins into the HVAC system. Poorly fitted natural gas lines can leak carbon monoxide into the living space. Ungrounded neutrals can upset the EMFs throughout the house. These are all mistakes that trades can make. Homeowners clearly benefit from minimizing the odds of these mistakes.

Basements minimize execution risk for trades when compared to other types of foundations. The jobs that trades do are hard enough. Imagine how hard it would be to install a whole HVAC system while crawling on your belly in a dark 2-foot tall crawlspace. Now, imagine how hard it would be to chisel the natural gas lines into a concrete foundation. Then, after installation, it is that much harder to detect problems and perform maintenance on those systems.

Accordingly, a properly constructed and maintained basement is the best foundation on which to construct and maintain a healthy home. New and existing basements will perform well as long as they are treated with the same care as other interior spaces in the home. In particular, basements should be conditioned (i.e. encapsulated, insulated, dehumidified, heated, and cooled) just like the rest of the home.

One exception is worth calling out. Some existing basements were not designed to be finished. These basements are generally not properly waterproofed. Finishing such basements can be a mold hazard. In other words, if the foundation was never properly waterproofed with the intention of supporting a finished basement, then covering those walls in insulation and drywall invites moisture penetration and water-related damage. Homeowners in this particular situation should consult with a licensed architect or engineer before finishing their basements.

A concrete slab on grade is a solid concrete block that sits above ground level. In my experience, a properly designed concrete slab has the best long-term mold avoidance performance. They are simply the least likely to suffer from moisture penetration in a way that facilitates mold growth. In general, a properly installed concrete slab on grade foundation is an inert rock that does not impact the indoor air quality of the home. However, a slab foundation makes plumbing, electrical, and HVAC more challenging. As I mentioned earlier, concrete slabs increase the execution risk of electrical and plumbing work. Worse yet, HVAC systems in concrete slab houses generally end up in attics. Locating HVAC equipment outside the conditioned envelope of the building creates mold and indoor air quality challenges that I would prefer to avoid. For these reasons, concrete slab on grade foundations is my second choice.

A crawlspace is typically a three-foot-tall cavity beneath the first floor of the home. Crawlspaces are a cheap and ergonomic building option in areas of the country where basements are difficult to build. In my experience, crawlspaces have the worst long-term performance in terms of resisting mold growth.

Crawlspaces are generally unencapsulated. Building codes will require unencapsulated crawlspaces to have numerous vents. The vents are generally passive. They are designed to allow the wind to dilute the air in the crawlspace. Unfortunately, these vents leave the space under the house completely exposed to the elements. Toxins and mold spores can blow in just as easily as fresh air. Worse yet, crawlspaces are generally designed to house HVAC equipment. Mold and toxin penetration in the HVAC equipment of unencapsulated crawlspaces is remarkably common.

In common unencapsulated crawlspaces, a large fraction of all the indoor air in the house enters through the crawlspace. In this situation, the fiberglass insulation commonly found on the underside of the first floor acts as a sort of filter. Inevitably, the insulation will collect dust and host mold growth, which can easily find its way into leaky ductwork.

Accordingly, I do not recommend crawlspaces for new construction. For existing construction, I recommend full encapsulation and climate control in the crawlspace. An encapsulation is a thick plastic vapor barrier (at least 15 mils) that isolates the crawlspace area from the outdoors. In turn, the encapsulation stops outdoor air and moisture from pushing on the subfloor of the house. I recommend climate controlling the crawlspace with insulation and dehumidification to reduce the risk of mold growth.

6.5 Radon and other soil gases

Indoor airborne radon is an easily solvable problem. Testing and remediation are simple. I recommend that everyone test for radon. I recommend that homeowners mitigate as necessary until they achieve an indoor radon reading below 1 pCi/L. I recommend testing only after construction has completed and the home has been conditioned to a livable temperature for at least two weeks.

Radon is a radioactive gas emitted from soil. It comes from naturally occurring uranium. Radon tends to be more common in the northern states, but it can occur anywhere. High radon is more likely to occur in homes with basements simply because basements are further underground.[9] (Basements are also more common in the north.) Naturally occurring uranium is an unpredictable and localized geological phenomenon. Accordingly, house-to-house variation in radon levels within the same neighborhood can be dramatic.

Radon is a carcinogen targeting the lungs.[10] The EPA recommendation for action is 4 pCi/L, which is equivalent to everyone in the home smoking 8 cigarettes per day. My lower action level comports with the advice of radon specialists. It also follows the theory anything carcinogenic at high levels is probably inflammatory at lower levels. The average outdoor radon level is 0.4 pCi/L.[3] Accordingly, I believe 1 pCi/L is achievable almost anywhere with an appropriate foundation exhaust fan (i.e. radon fan) and good ventilation (i.e. the "V" in HVAC).

A radon test costs about $20. Roughly two out every five homes will need mitigation to achieve a reading below 1 pCi/L.[7,6] Mitigation via the installation of a foundation exhaust fan by a certified professional is approximately $2,500 depending on the area. A foundation exhaust fan will provide some protection from other harmful soil gas exposure as well. A thick vapor barrier (15-mil or greater) under the slab provides extra protection.

All of these measures serve to block soil gas intrusion. Radon is just one high-profile soil gas. Other soil gases are generally man-made pollutants that have settled into the earth and groundwater. Water is also a soil gas. As such, these recommendations help reduce humidity from entering the home through the foundation, basement, and crawlspace. Moisture intrusion from these sources is particularly hazardous. Moisture tends to carry the chemical footprint of its source. Accordingly, reducing subsurface moisture intrusion makes it harder for subsurface toxins to get into the home.

References

[1] American Lung Association. "Carpets and Rugs". In: (2026). URL: https://www.lung.org/clean-air/indoor-air/indoor-air-pollutants/carpets.

[2] Environmental Protection Agency. "Carpet and Indoor Air Quality". In: (1992).

[3] Environmental Protection Agency. "What is the average level of radon found in homes in the U.S.?" In: (2026). URL: https://www.epa.gov/radon/what-average-level-radon-found-homes-us.

[4] Sarah R. Haines et al. "Ten questions concerning the implications of carpet on indoor chemistry and microbiology". In: *Building and Environment* 170 (Mar. 2020), p. 106589. ISSN: 0360-1323. DOI: 10.1016/j.buildenv.2019.106589.

[5] J. William Hirzy and Rufus Morison. "Carpet/4-Phenylcyclohexan Toxicity: The EPA Headquarters Case". In: *National Federaion of Federal Employees* (1991).

[6] Longxiang Li et al. "High-resolution national radon maps based on massive indoor measurements in the United States". In: *Proceedings of the National Academy of Sciences* 122.3 (Jan. 2025). ISSN: 1091-6490. DOI: 10.1073/pnas.2408084121.

[7] F. Marcinowski. "Nationwide Survey of Residential Radon Levels in the US". In: *Radiation Protection Dosimetry* 45.1-4 (Dec. 1992), pp. 419–424. ISSN: 1742-3406. DOI: 10.1093/rpd/45.1-4.419.

[8] Wayne R. Ott, Anne C. Steinemann, and Lance A. Wallace. *Exposure Analysis*. CRC Press, 2007.

[9] Fintan K. T. Stanley et al. “Radon exposure is rising steadily within the modern North American residential environment, and is increasingly uniform across seasons”. In: *Scientific Reports* 9.1 (Dec. 2019). ISSN: 2045-2322. DOI: 10.1038/s41598-019-54891-8.

[10] World Health Organization. “Radon”. In: (2023). URL: https://www.who.int/news-room/fact-sheets/detail/radon-and-health.

Chapter 7

Walls, Roofs, and Insulation

A stick frame wall assembly is like a sandwich of materials layered on top of one another. I will list eight different materials that should create the wall of a traditional stick frame house. The stick frame is a staple of the residential construction industry in America. For financial reasons alone, the vast majority of Americans will need to consider some form of stick frame construction. I will describe in detail my version of a sensible stick frame wall system in this chapter. Many premium alternative wall systems exist, and many of them are highly attractive from an indoor air quality perspective. I will give them a light treatment in this chapter.

7.1 Executive Summary

For existing builds, I recommend homeowners leave their walls, roofs, and insulation in-place unless there exists a compelling reason to replace them.

For new construction, I will make a sensible default recommendation and a premium recommendation. As a sensible default, I recommend a slightly modified standard stick frame wall assembly. In order, from inside to outside, I recommend pure mineral paint over (2) paperless drywall with a (3) smart membrane beneath it. Then, I recommend (4) Knauf fiberglass insulation in front of a (5) non-foam exterior plywood protected by (6) a water-resistant barrier ("WRB"). Then, I recommend (7) a rainscreen atop the WRB and behind (8) any exterior siding. For a premium alternative, I recommend a uniform whole-wall organic wall assembly like Faswall or rammed earth.

7.2 Background

My recommendations with regards to wall assemblies and roofing are pretty standard. A typical builder following this guidance will only have to make a few specific material substitutions. Additionally, he may have to use only one or two unfamiliar materials. Otherwise, my wall assembly and roofing recommendations are basic. This is an intentional feature of my guidance. Walls and roofs are some of the most labor-intensive parts of home construction. The U.S. building industry has a very specific template for walls and roofs that is a product of its building codes and building culture. We as reasonable homeowners are fortunate that we can achieve fairly non-toxic walls and roofs by simply substituting in our favorite flavors of the same core materials.

There are innovative alternative building systems out there. I will touch on them. But, they are so varied in both region and cost that they do not make sense to recommend as a one-size-fits-all solution. Most of these innovative solutions are new takes on old concepts. For example, engineered concretes (like Faswall and Hempcrete) and rammed earths (like Sirewall) provide uniform end-to-end alternatives to common wall assemblies. Additionally, mineral wool insulation and some borate-fortified natural insulations (like cellulose, wool, and wood fiber) are toxicologically competitive with untreated fiberglass.

7.3 Encapsulation

Most wall assemblies are only a cause of concern for homeowners if they are contaminated in some way by mold or animal infestation. Undamaged wall assemblies benefit from encapsulation. The term *encapsulation* refers to the fact that many of the potentially toxic chemicals in materials like the insulation or the sheathing are enclosed in solid materials like siding and drywall. Encapsulation significantly reduces toxicity risk to homeowners by (1) preventing occupants from physically contacting the surface, (2) blocking direct interface between the material and the indoor air, and (3) somewhat containing unbound toxins. Accordingly, homeowners only marginally benefit from selecting materials with highly pure ingredients when considering insulation, sheathing, siding, and roofing. The principle concerns of material selection when it comes to wall assemblies and roofs are (1) the prevention of mold growth and (2) the avoidance of specific highly hazardous materials. I will proceed with a discussion of material selection for each of the components of a wall assembly and a roof. (See Chapter 4 for a discussion of paint, which I will not specifically address here.)

7.4 The Sensible Default Wall Assembly

I recommend paperless drywall everywhere. The only significant toxicity risk inherent to drywall is the paper that coats it. The paper layer on top of the the drywall is feedstock for mold in the event of moisture exposure. Builders are somewhat used to using paperless drywall in bathrooms where high humidity is normal and moisture exposure is expected. Paperless drywall uses a fiberglass mesh instead of paper, rendering it virtually mold-proof. The fiberglass mesh makes it a little heavier and a little more difficult to paint, as it requires a skim-coat of plaster to be smooth.

Ideally, a smart membrane would be situated in between the insulation and the drywall. The smart membrane is a plastic-like sheet through which water can only pass in one direction. The smart membrane should be oriented such that moisture can pass from the insulation, through the drywall, and out into the interior of the home.

Beneath the smart membrane, inside the stud cavity, lies the insulation. I recommend a Knauf fiberglass blown-in-blanket system. Fiberglass is a completely inert material that is theoretically perfectly safe to homeowners. There are many fiberglass insulations on the market that use a long list of biocides, flame retardants, and binders we seek to avoid. My particular favorite, Knauf, is very minimalist. It lists only fiberglass and mineral oil as ingredients. (See Declare List ID: KNF-0007.) Mineral oil is a petroleum byproduct that is fairly inert when left undisturbed. Mineral oil seems to be the only concerning (but necessary) manufacturing input to fiberglass insulation. Accordingly, I recommend this Knauf product because, ingredient-wise, it is the simplest fiberglass insulation known to me.

Homeowners that seek to avoid mineral oil can substitute in other natural insulations. Alternative insulations are, by and large, a grand theater of adverse substitution. Many alternative insulations are more focused on recyclability and greenness than health. Accordingly, they often use recyclable materials as substrates but add toxic binders, biocides, and flame retardants to achieve the performance normally expected of fiberglass.

Almost all alternative insulations need some form of chemical additive to meet code requirements for fire resistance. My preference is disodium octaborate tetrahydrate (or "D.O.T."). Pure boron-based salts like D.O.T. are generally low-toxicity and not often triggering to chemically sensitive individuals. They are effective at penetrating porous substrates (like cellulose and wood) because of their high water-solubility. Their water-solubility is desirable in the manufacturing phase but not in the use phase. Excessive moisture can cause leakage and depletion of boron salts. It is not uncommon to find crystallized boron salts at the bases of walls by moldings. As long as this risk is managed effectively through effective climate control, boron-based salts are the best code-compliant flame retardant for alternative insulations.

Many proprietary boron-based flame retardants attempt to improve D.O.T.'s unattractive solubility properties with proprietary additives. Historically, some of these additives have been dangerous. Some that are currently in use are not well studied. Accordingly, I recommend avoiding proprietary boron-based flame retardants and instead relying on products that transparently use D.O.T. I am aware of a few cellulose and wood fiber products that meet this criteria, but I'll defer interested homeowners to the most up-to-date Declare List entries.

Mineral wool insulation is made from basalt rock that is melted and spun into fibers, along with a phenol-formaldehyde binder. Generally, mineral wool is naturally fire resistant, has a better insulative value and does not hold on to moisture like many other insulation materials can. Provided that my sensible wall assembly is followed, moisture retention in the wall should not be a concern no matter which insulation you go with. For pricing comparison, mineral wool is roughly twice the cost of fiberglass. For those chasing energy performance numbers or if you prefer superior sound deadening characteristics, mineral wool is a great choice.

Behind the insulation is the exterior sheathing. See Chapter 5 for details on engineered woods to inform your choice of exterior sheathing. I recommend high-quality engineered woods for exterior sheathing and subfloors.

On the exterior-side of the sheathing, I recommend a weather-resistive barrier ("WRB"). These sorts of barriers prevent moisture intrusion while allowing water vapor to exit from underneath.

I recommend one of the following solutions for a WRB. The Huber Zip System is an OSB sheathing that comes with a factory-applied WRB. The Zip System uses OSB and meets my definition of high-quality engineered wood as described in Chapter 5. Homeowners specifically seeking to use a different brand of sheathing are advised to use stick-on WRBs like Henry Blueskin or SIGA Majvest SA.

I recommend avoiding staple-on WRBs. Staple-on WRBs, as commonly applied, perform poorly against wind-driven rain and snow. This sort of moisture intrusion can be a contributing factor to future mold growth within the wall assembly years down the road.

On top of the WRB, I recommend a rainscreen. A rainscreen is a textured surface that allows condensation to gracefully exit from underneath the exterior sidings. And, finally, your choice of exterior siding will be the last element of the wall assembly.

My default recommendation for a wall assembly is pretty basic. This is intentional. I do not want to reinvent this process that is already so well-understood by builders across the U.S., especially when it is sufficient for the vast majority of chemically sensitive homeowners. Naturally, as we move farther and farther along the wall assembly and away from the interior of the home, we are less concerned about the toxicity of the materials chosen. Accordingly, I give almost no guidance to offer as to the choice of a WRB, a rainscreen, or exterior siding. As long as my other recommendations are followed, materials choices on these components are very unlikely to affect indoor air quality.

7.5 The Sensible Default Roof

In regards to indoor air quality, the materials selected for the roof are of minimal importance. Standard asphalt shingles are fine. Most premium alternatives are also fine. For roof insulation, homeowners should follow the guidance in the above section on wall insulation. Some key materials to avoid will have particular importance to roofing. I will discuss these in the next section.

Mainly, roofs need to resist moisture penetration, and attics need to be kept clean. Accordingly, I will avoid a specific discussion of roofs. My guidance about resisting moisture penetration will apply to all areas of the house, including roofs. Further, my guidance about keeping exterior interfaces clean will apply to attached garages, unfinished basements, crawlspaces, and attics equally. I will address these issues all together in a separate part of the book.

7.6 Spray foam insulation

I recommend against spray foam insulation. Foam insulation is becoming increasingly popular for its versatility and performance. Foam insulation is a diisocyanate-based compound that is cured on site. It can be sprayed as a gelatinous beam into any unoccupied space. Then, it will expand and harden to form a solid mass of high-performance insulation. Foam insulation generally comes in a kit with two containers that resemble propane tanks. Each container holds a reactant. An applicator hose combines the reactants in the appropriate ratio as they spray. The reactants combine within the spray beam and cure on the surface.

My position on foam insulation is nuanced. I believe that diisocyanate-based products (like pMDI discussed in Chapter 5) are safe and valuable when manufactured properly. I believe the only way to properly manufacture diisocyanate-based products is off-site (i.e.

away from the home in a controlled manufacturing facility). This belief derives from the possibility that unreacted reactants, including diisocyanate, can off-gas from cure-on-site products and harm occupants. In other words, any *incomplete cure* poses a hazard to occupants.

In my experience, cure-off-site diisocyanate-based resins like pMDI have the lowest risk of harmful off-gassing. Typically, the risk of off-gassing from this sort of product is negligible. Accordingly, I highly recommend pMDI-based engineered woods in Chapter 5. pMDI-based engineered wood is an example of the appropriate usage of diisocyanates.

Publicly available research and my professional experience suggests that foam-based diisocyanates are significantly riskier than resin-based diisocyanates. For example, research into cheaply manufactured polyurethane-based foams (another diisocyanate product) suggests that they contain significant unreacted quantities of diisocyanates. This research previously fueled significant concern that polyurethane foam pillows were hazardous.[1]

Cure-on-site diisocyanate products are significantly riskier than their cure-off-site counterparts. Cure-on-site products are more likely to incompletely cure due to (1) installer error, (2) environmental variability, and (3) the absence of post-install quality controls. Any deviation from the manufacturer-supplied guidance in the application of cure-on-site products can result in an incomplete cure. For example, with regards to foam insulation, if the tanks are not precisely pressurized such that they emit that exact intended quantities of reactants, unreacted chemicals will remain. My professional experience suggests that tank lines often experience partial or complete clogs that contribute to incomplete cures. Environmental variability can exacerbate this problem, as well. Differences in the way the reactants in each tank respond to changes in temperature can force tank pressures into disequilibrium. Then, environmental deviations in the application site can further exacerbate this problem. High humidity and extreme temperatures can cause an incomplete cure. Further, inappropriate application of subsequent layers of foam can cause an incomplete cure.

It is no secret to the manufacturers or the installers that uncured reactants are dangerous. Spray foam application requires significant safety precautions from the installers, including full-body suits and masks with continuous fresh air supplies. (My professional experience suggests that installers often rely on less than ideal protective measures.)

The energy policy benefits of spray foam insulation explain why it is so common. Spray foams are the highest-performing commercially available insulation choice for residential homes. In other words, the R-value of spray foam in a 2x4 stud frame is the highest on the market. Many state building codes have been incrementally increasing R-value requirements for new home construction to such a level as to essentially mandate spray foam. In these states with cold climates and aggressive energy efficiency goals, spray foam quickly becomes the most economically viable option for common stick frame construction. Certain non-traditional and more expensive construction methods (like deep stud cavities and solid foam continuous insulation) may be viable in these states. In any case, the average new construction in such a state will leverage spray foam.

Spray foam has an additional beneficial side effect that makes it attractive to energy-conscious builders. Spray foam is the only insulating material that creates an air seal. Imagine, for example, a PVC pipe or an electrical cable (or any utility) running through a hole in a stud cavity. Every utility hole on the top, bottom, or side of that stud cavity is a place for outside air to penetrate the building envelope. Normally, a builder might throw some fiberglass batts on top of the stud cavity then cover the stud cavity with drywall, leaving the tiny air gaps on the edges of the utilities unaddressed. Other builders interested in air sealing might use some other product to seal up the air gaps. Spray foam users can insulate the wall cavity and close up the air gaps at the same time. The spray foam grows and expands a little upon application. As it does, it finds these air gaps, crawls into them, and hardens to create an air seal. I obviously strongly advise against the use of spray foams. Regardless, it is helpful to understand the mindset of builders that use them.

Many states with progressive building codes are mandating blower door tests (see the HVAC chapter) at certain performance levels (below 3 ACH) for a house to pass inspection. This requirement is intended to make homes more energy efficient. I have heard some home inspectors say that the sum of all air gaps in the house must be less than 36 square inches in order to get a blower door test below 3 ACH. Builders have reported to me that spray foam insulation is helping them pass blower door tests by sealing up air gaps. Many of these same states are requiring very high insulation performance (expressed in high R-values). Many builders find that the only way to meet both requirements simultaneously is through the use of spray foam insulation. Avoiding spray foam in these jurisdictions is necessarily more expensive.

References

[1] Cheryl A. Krone and Tom D. Klingner. "Isocyanates, polyurethane and childhood asthma". In: *Pediatric Allergy and Immunology* 16.5 (Aug. 2005), pp. 368–379. ISSN: 1399-3038. DOI: 10.1111/j.1399-3038.2005.00295.x.

Chapter 8

HVAC

HVAC is an acronym for *Heating, Ventilation, and Air Conditioning*. The HVAC system is the biggest (and generally the most expensive) appliance in a house. Unfortunately, there are a dizzying number of ways for homeowners to spend money on their HVAC systems. As always, this chapter will attempt to help readers identify high-impact HVAC issues and prioritize HVAC investments.

8.1 Background

The relationship between the HVAC system and the house is like the relationship between the circulatory system and the human body. The parts must be clean and efficient in order for the whole to be healthy and functional. Additionally, the system is cyclical. A failure of a part will impair the performance of the whole. (That is about as far as I will stretch this metaphor.)

Designing an HVAC system is a fairly opinionated engineering exercise. HVAC design is an engineering exercise in that every design decision is a calculated trade-off. HVAC design is opinionated in that the objectives of the HVAC system can be defined and balanced according to homeowner preference. Possible objectives include (1) energy efficiency, (2) indoor air quality, (3) cooling and heating performance, (4) filtration performance, (5) ventilation performance, (6) space efficiency, (7) cost, (8) durability, (9) safety and (10) code compliance. To complicate matters further, HVAC installers are heavily constrained with regards to product selection. The HVAC installers have near-exclusive relationships with specific OEMs. And, the OEMs must follow ever-evolving federal and state energy regulations.

In consideration of all of the above, the homeowner is left with few meaningful options for success. The system that a homeowner bought 10 or 20 years ago is probably no longer legal to sell and install. The embedded infrastructure (i.e. supplies, returns, and ducting) is almost certainly too expensive to modify. Mechanical engineering services are useful but expensive. Worst yet, the vast majority of HVAC decisions are made in a hurry. Most HVAC installations in existing homes are performed on a rushed timeline in response to a total system failure. Homeowners and installers rarely have adequate time to make carefully considered HVAC decisions.

The frustration homeowners feel when purchasing and maintaining HVAC systems is similar to that of buying a car. Both assets are highly regulated. Both have highly specific supply chains. Both offer the purchaser a dizzying array of choices to make. Still, the purchaser ultimately feels like she has very little meaningful autonomy. In both cases, the savvy consumer can at least maximize the safe and useful life of that asset. This chapter will give homeowners clear guidance for designing and maintaining an HVAC system that prioritizes indoor air quality without ignoring other objectives.

8.2 Definitions

The following discussions will assume a lot of knowledge about HVAC design and building science concepts. The following definitions should help readers interacting with these concepts for the first time.

Returns bring air into the HVAC system. These typically appear in walls and ceilings. They are often larger than supplies. *Supplies* bring conditioned air from the HVAC system to the living space. They are typically more numerous than returns. They can often have adjustable dampers. They can be in floors, walls, and ceilings. *Plenums* are the central distribution points of the HVAC system. A plenum consolidates air from the ductwork into a central location before passing it to the HVAC systems for conditioning. Then, another plenum distributes conditioned air back through the supply ducts. Plenums are often empty metal boxes, but they can also be wooden floor cavities. The *ducts* are the tendrils that connect the plenums to the supplies and returns.

CFM stands for *cubic feet per minute.* It is a measure of the volume of air moving through a point (typically a door, window, or fan). The power of a fan is typically measured in CFM. *ACH* is the *air changes per hour.* It measures the same fundamental information as CFM, except it takes into account the volume of the container the air is flowing through. For a container with size in cubic feet, V, the follow conversion holds.

$$\text{ACH} = \frac{60 * \text{CFM}}{\text{V}}$$

The ACH50 is the ACH measured at a relative interior pressure of 50 Pascals during a standard blower door test. The *blower door test* standardizes the relative interior pressure to 50 Pascals before measurement in order discern a uniform notion of ACH. Blower door tests are a diagnostic procedure for measuring ACH50 in a house. A blower door test involves hooking up a large faux door with fans and meters in it to the front door of the house.

Manuals J, S, T, and D are metholodogies for HVAC system design published by the Air Conditioning Contractors of America ("ACCA"). Each manual is responsible for a separate aspect of HVAC design. J is for load calculations. S is for equipment sizing. T is for supply and return placement. D is for duct sizing. All four manuals are relevant during initial construction. Only manuals J and S are relevant during HVAC system replacements. The Manual J calculation is notoriously laborious but valuable.

An *MEP engineer* is a licensed professional engineer specializing in the *Mechanical, Electrical, and Plumbing* systems of a building. MEP engineers can generally provide a more sophisticated and comprehensive design perspective than an HVAC contractor. They are commonly retained on complex or commercial projects during new construction. On simple residential projects and HVAC system replacements, the HVAC contractor typically fills the equivalent design role.

ERV is an acronym for *energy recovery ventilator*. ERVs are designed to efficiently increase ventilation while maintaining humidity. Ventilation is the "V" in "HVAC". It refers to the flow of fresh outdoor air into the house. ERVs address the problem of insufficient ventilation in tight energy-efficient homes. HRVs are closely related, except they only exchange the temperature of the air, not the temperature and humidity that an ERV does. For the purposes of this discussion, I will only refer to ERV..

8.3 Executive summary

The executive summary for this chapter is broken up into subsections due to its length.

8.3.1 Interior air pressure

Negative relative interior air pressure should be avoided wherever reasonable, both on a house-wide basis and on a room-by-room basis. Any exhaust activity above 50 CFM for 30 minutes requires compensating safe air intake to avoid negative relative interior air pressure. Any bathroom exhaust fans with unknown capacity should be assumed to be 50 CFM each for the purpose of this rule. Any kitchen range hoods with unknown capacity should be assumed to be 300 CFM for the purpose of this rule. For new construction, I recommend integrating the ERV with the exhaust systems to equalize the interior air pressure with the exterior air pressure. (Slightly positive relative interior air pressure is desirable but difficult to achieve.) For existing construction, I recommend hiring an HVAC contractor to install an ERV to safely compensate for unintended pressure loss. My next best recommendation is to run conditioned makeup air through the HVAC. For existing constructions not currently adhering to this guidance, I recommend running bathroom exhaust fans either (1) one at a time for 30 minutes or less or (2) with a window open. For existing constructions not currently adhering to this guidance, I recommend opening a window whenever the range hood is running.

In new construction, I recommend that every closeable room above 100 square feet has both a supply and a return. I will define a closeable room as one whose airflow can be choked off by closing its interior doors. I will define any home that does not meet this design criteria as *internally unbalanced*. In existing internally unbalanced homes, I recommend that homeowners leave interior doors open as much and as long as possible. I do not recommend renovating ductwork to address internal imbalance in existing homes.

8.3.2 Leak and pressure testing

Testing ductwork for leaks is advisable only if there exists other evidence of the intrusion of toxins into the ductwork. This generally requires an identifiable toxin to exist (1) in an attic that houses ductwork, (2) in a crawlspace that houses ductwork, or (3) in the ductwork itself.

A blower door test is advisable only in one of the two following circumstances. (1) It is required by code. (2) The leakiness of the house is expected to be a meaningful factor in the design of the HVAC system. In houses over 20 years old, the leakiness value can be assumed to be 5 ACH50 (rather than measured through a blower door test) for the purposes of sizing calculations.

8.3.3 System sizing

In new construction, I recommend that a competent professional (either a specially certified HVAC technician or an MEP engineer) perform manual J, S, T and D calculations to properly size the HVAC system. For existing constructions evaluating HVAC system replacements and upgrades, I do not recommend all of these calculations. Rather, in existing constructions, I recommend performing the manual J and S, as well as solving for specific HVAC deficiencies with compensating upgrades.

8.3.4 Ducting materials

In new construction, I recommend installing rigid metal ducts. In existing constructions, I do not recommend replacing flex ducts unless they are both old (over 15 years) and uncleanable. Here, an *uncleanable* flex duct is defined as one with microbial growth that cannot be a remedied by proper filtration, purification, and duct cleaning. For all homes, I recommend regular duct cleanings (upon move-in and then every 3 to 5 years).

In existing constructions, any and all fiberglass duct board should be replaced, preferably with rigid metal ducts. Fiberglass duct board is fragile, highly textured, and impossible to clean. Accordingly, fiberglass duct board carries an unacceptable toxic air intrusion risk and unacceptable mold growth risk.

8.3.5 Purification and dehumidification

For existing construction, I recommend that the HVAC system is upgraded to support the Dynamic Air RS4 with a UV-C bulb (not the UV-V "activated oxygen" bulb). Prior to this recommended upgrade, I recommend that homeowners use the supply and return air filters that the HVAC system was designed against. I recommend that homeowners in this situation replace the air filter along the most frequent manufacturer-recommend guidelines (typically monthly for weaker filters). I specifically advise against upgrading the air filters in the home without professional assistance to avoid damaging the HVAC system. (Stronger filtration media may require more power than an HVAC system has to offer.)

For new construction, I recommend designing the HVAC system to support MERV16 filtration. Specifically, I recommend the HomeCleanse Shield. Further, I do not recommend UV-C for new construction. I believe healthfully designed homes do not need supplemental electronic purification like UV-C. In all cases, I recommend avoiding UV-V lights, ionization-based air cleaners, and ozone-based air cleaners unless under my specific guidelines.

I encourage the strategic deployment of standalone air purifiers and standalone dehumidifiers for existing construction. I encourage the usage of HEPA filters and activated charcoal in standalone air filters. I encourage the usage of standalone dehumidifiers in places that need protection from severe humidity swings like crawlspaces, basements, and bathrooms.

8.3.6 Fuel type

For new construction, I have no particular preference for gas or electric heaters as long as the heater uses forced air. For gas heaters, I recommend fan-blown exhaust rather than passive exhaust. For electric heaters, I remind readers to consult the EMF safety guidance related to major appliances in Chapter 9. For existing construction, I do not recommend modifying the location of major HVAC appliances. If testing discovers natural gas combustion byproducts in the indoor air, I recommend upgrading any passive exhaust to fan-blown exhaust for gas heaters.

8.3.7 Appliance location

For new construction, I recommend that major HVAC appliances are located in the conditioned space wherever reasonable and allowed by code. I define the *conditioned space* to be the parts of the house that are subject to heating, cooling, insulation, dehumidification, filtration, and encapsulation. The most straightforward way to place HVAC appliances in the conditioned space is to place the relevant pieces in utility closets that are contiguous with the living space. For existing homes where this advice is not followed, I recommend encapsulating, insulating, and dehumidifying any spaces that house HVAC equipment to the maximum reasonable extent. In particular, I recommend encapsulating and dehumidifying any crawlspaces. For attics containing HVAC equipment, I recommend insulating and taping the HVAC equipment itself, especially the plenum box.

8.3.8 Mini-split systems

I recommend avoiding mini-split systems. Mini-split systems (a) eliminate the possibility of whole-home dehumidification, (b) eliminate the possibility of whole-home air purification, and (c) tend to produce hazardous condensation through rapid localized temperature swings.

8.3.9 Joist panning

I recommend eliminating any and all instances of joist panning in both new and existing construction. *Joist panning* refers to the practice of using sections of subfloor in place of ductwork to direct airflow. I recommend any instances of joist panning to be replaced with rigid metal ducting. Joist panning is an obvious and easily avoidable mold risk.

8.4 Incorrectly sized HVAC systems

I believe mold exposure is the most prescient house-related threat to occupant health. Incorrectly sized HVAC systems can encourage mold growth by failing to properly regulate humidity. First, I will explain how HVAC systems are supposed to regulate humidity. Then, I will explain how improperly sized systems (too large or too small) can lead to excess humidity.

Total humidity is a measure of the number of water granules in the air. Relative humidity is the total humidity divided by the maximum theoretical total humidity achievable at the current temperature. If water is injected into the air when the relative humidity is already 100%, condensation will form on surfaces, creating a mold hazard. Hotter temperatures have a higher capacity for total humidity. Accordingly, as temperature drops while total humidity is held constant, the relative humidity can reach 100%, which creates condensation, which creates a mold hazard.

Additionally, not all surfaces in the house are the same temperature. Given some ambient air temperature and some total humidity level, there exists a temperature called the *dew point* below which condensation will deposit on hard surfaces. This form of condensation is a mold hazard both to the cold surface and to any nearby surfaces that condensation might drip onto.

In general, the winters are cold and arid, and the summers are hot and humid. Of course, the specifics vary dramatically by region of the country, but the general principle holds. In general, homeowners run the heat in the winter to warm up the house and the air conditioning in the summer to cool down the house. Heating and air conditioning work very differently. They operate on virtually independent systems. Still, they both serve to reduce the humidity in the house in some way.

Heating works by natural gas combustion or electric induction. The air is heated as it flows over a coil as it circulates through the house. Heating the house reduces the relative humidity by raising the temperature. The goal of heating is not to dehumidify the house, because the conditions in which homeowners use heat are generally already dry. The reduction in relative humidity is merely a side effect. While very low humidity is unhealthy, dry indoor air during the wintertime is not usually a health hazard. And, it is certainly not a mold hazard. Typically, the normal household activities of the occupants sufficiently offset the seasonally low humidity. Normal activities that generate humidity include cooking, bathing, and breathing.

The safety of the home depends on the air conditioning system's ability to properly dehumidify the indoor air. Since mold exposure is the most prescient house-related threat, high humidity is a much greater threat than low humidity. Accordingly, the remainder of this discussion about right-sizing HVAC systems will focus on the effects of improperly sized air conditioners.

The air conditioning system passively dehumidifies the air as it runs. This is unintuitive but important. Readers only need to understand the parameters of the following engineering problem. They do not necessarily need to understand exactly how air conditioning or the refrigeration cycle works. In short, air conditioning works by running air over a cold coil (the evaporator coil). The evaporator coil is cooled by the refrigeration cycle. The refrigeration cycle cools the coil by using mechanical energy to repeatedly evaporate a refrigerant solution from a liquid state to a gaseous state. As the ambient air from the house runs over the very cold coil, the moisture from the ambient air condensates on the coil. The condensation drips into a drain pan and flows out of the house via the condensate drain line. Accordingly, air conditioners both decrease the air temperature and decrease the total humidity. It is important to recognize that air conditioners decrease the total humidity (rather than strictly the relative humidity) by directly removing water from the air. Information about the refrigeration cycle is rudimentary and easy to access elsewhere. Again, it is not important for readers to understand the refrigeration cycle itself. Rather, it is important that they understand the parameters of this engineering problem.

In rough terms, the cooling capacity per unit of time of the system is proportional to its size (i.e. the tonnage), but the dehumidification capacity per unit of time of the system is a constant.

Therefore, we as homeowners can cool a house faster with a bigger air conditioner, but we cannot dehumidify a house faster with a bigger air conditioner. Similarly, a smaller system will require more time to completely cool a house, but will perform more dehumidification by running longer.

Now, consider how a typical thermostat works. A thermostat set to cooling mode will measure the temperature in the house. Then, it will turn on the air conditioner if the the temperature in the house is above the specified target. Then, the thermostat will continuously monitor the temperature in the house until the temperature reaches the target on the thermostat. When the temperature reaches the target, the thermostat will turn off the air conditioner. If this sounds rudimentary, that is because it is. The simplicity of thermostat design is actually a really big problem.

Homeowners depend on their air conditioning systems to perform two functions (temperature reduction and total humidity reduction). The thermostat is only measuring progress towards one of these goals (temperature reduction). The whole ecosystem of commercially available HVAC technologies and products more or less assumes that enough cooling corresponds to enough dehumidification. This assumption is true enough for a properly sized system. Improperly sized systems violate this assumption, and the consequences imperil occupant health.

An oversized air conditioner will cool the house (i.e. reach the thermostat temperature) too quickly to allow for sufficient dehumidification. This creates excessive humidity in the home (above 60% relative humidity), which is a mold growth hazard. If the corresponding excessive humidity does not cause mold growth directly, other common HVAC issues can compound this problem to create an indoor environment extremely conducive to mold growth.

An undersized air conditioner will never successfully cool the house. Excessively hot homes (above 78 degrees Fahrenheit) have high total humidity capacity, which is conducive to mold growth. Additionally, undersized systems will be continuously overworked, so they will fail early. The inevitable early failure event in the undersized system will create a period of potentially days to weeks, typically during a hot and humid time of year, where air conditioning and dehumidification are completely unavailable. This event will likely allow mold to entrench itself in the home.

To properly size an HVAC system for new construction, a certified HVAC professional or an MEP engineer should run Manual J, D, S, and T calculations prior to system selection and initial installation. For existing constructions, at the time of replacement, if a unit with the exact same specifications as the existing unit is not available, the Manual J and S calculations should be run again to select a new system. I do not recommend the Manual D and T for existing, due to cost considerations and home disruption. These approved calculations consider all of the necessary variables to properly size an HVAC system, including the number, size, and positioning of rooms, floors, and windows. These calculations even take into account the impact of the occupants on the energy requirements of the system.

There exist numerous methods of sizing HVAC systems. They are all just different mathematical models. The simplest mathematical model tries to correlate the size of the house (often in square feet) to the power of the system (in tonnage). This model has only one input variable (the square footage). It is simply not complex enough. More than a few variables are required to properly design an HVAC system. Other models of intermediate complexity exist. The Manual J model is, if anything, a little too complex. Maybe it requires just a few too many variables. Accordingly, some of the Manual J variables can reasonably use assumptions or estimates (e.g. the blower door values for older homes can be assumed or estimated). In any case, the Manual J, Manual D, and Manual S calculations are very standardized and well-known. Manual T is not as common in residential work, but it deals with grill sizing, air velocity and throw patterns. Any good faith effort by a competent professional to perform the calculations will provide high-quality data with which to design an HVAC system.

8.5 Negative pressure

Formally, negative relative interior air pressure (hereafter "negative pressure") refers to a state of the home where the interior air pressure is lower than the exterior air pressure. Negative pressure, both on a house-wide and on a room-by-room basis, should be avoided wherever reasonably possible. Negative pressure is a toxin intrusion mechanism at any level and a moisture intrusion mechanism at moderate to high levels. This section will attempt to imbue the reader with an intuition about interior pressure so she can deploy the right mix of lifestyle habits and home upgrades to prevent negative pressure.

Consider the following thought experiment. Imagine that you remove your front door and replace it with a giant spinning electrical fan. Imagine that the fan is really big and powerful. Imagine that it spans the frame of the front door top-to-bottom and side-to-side. You point the fan into the house, such that it moves air from the outside to the inside, and you turn it on. As you stand in the foyer of the house right in front the fan, you feel the intense airflow on your body. Still, the house does not blow up like a balloon. It does not explode like you are in an action movie. The house stays still. It remains rigid. A few things follow. The air pressure in the house rises. As the air pressure in the house rises, the pressurized air looks for ways to escape the house. The pressurized air escapes the house in every conceivable way. It flows through cracks in the walls, through the recess light sockets, through the electrical sockets, through the attic door, through any leaky windows, and so on. Eventually, as the fan continues to run, the pressure in the house reaches an equilibrium point. The pressure in the house rises to a high enough level to force air out of the cracks and crevices at the same rate that the fan brings air in. In a leaky old house, the pressure would be pretty low. In a tight energy-efficient house, the pressure would be pretty high. In both cases, the house is under positive pressure relative to the outdoors.

Now, imagine that you turn the fan around, such that the fan blew air from the indoors to the outdoors. You achieve the exact opposite effect. The house does not deflate like a balloon. Rather, it leaches air in from every crack and crevice in order to supply the fan with air. When the air pressure reaches equilibrium in the house, there is a constant inflow of air from all of the cracks and crevices in the house into the indoor air (and then out the front door). The house is under negative pressure relative to the outdoors.

Now, imagine that you take the fan off of the frame of the front door and place it in the middle of a football field. You turn the fan on and stand in front of it. You instantly notice that it is much more powerful than it was when it was pointed into or out of the house. That is because air is a fluid, and the fluid generates friction when it is dragged in and through and across surfaces. In this football field, there is almost no friction except the air acting on itself. There are no tight cracks to squeeze through. There is no net pressurization except for right in front of and right behind the fan. Accordingly, the fan blows much more powerfully in the middle of the football field. We can now exit the thought experiment.

If this hypothetical fan was rated at 50 CFM (cubic feet per minute), it would be the average bathroom exhaust fan. If this hypothetical fan was rated at 300 CFM, it would be the average kitchen range hood. At 1,200 CFM, it would be a severely overpowered kitchen range hood.

All of these devices are exhausting devices. They move air from the inside to the outside of the home. All of these devices can create negative pressure zones if they are run long enough to reach equilibrium. The smaller the enclosed space and the higher the CFM, the faster they reach equilibrium. For example, a bathroom exhaust fan might take 5 minutes to reach a negative pressure equilibrium when the bathroom door is closed, but 50 minutes when the door is open. If the bathroom door is closed, the loose electrical sockets compete with the underside of the bathroom door to supply air to the exhaust fan. If the bathroom door is open, the exhaust fan will take its time depressurizing the entire house before it requires additional air intake from cracks and crevices. Additionally, just like our hypothetical fan in the football field, the exhaust fan will move more air if the bathroom door is left open. The goal of a bathroom exhaust fan is to remove unpleasant smells. The goal of a shower exhaust fan is to remove steam. Both of these tasks involve removing a fixed quantity of air, hopefully as quickly as possible. Leaving the doors open not only avoids negative pressure in the bathroom, it also shortens the time required for the fan to do its job.

The HVAC system itself can create negative pressure zones if not carefully designed and managed. For new houses, I recommend that every closeable room above 100 square feet has both a supply and a return. This recommendation is designed to avoid the accidental creation of negative pressure zones.I call homes that do not follow this design criteria *internally unbalanced* with regards to the HVAC design. Most existing residential homes are internally unbalanced according to this definition. In such homes, simply closing the wrong door at the wrong time can interfere with the proper functioning of the HVAC system and create dangerous negative pressure zones.

Consider the following example to demonstrate this principle. Say that a typical U.S. home is 2,400 square feet with two floors. A common HVAC system in this home might have one supply in the downstairs hallway, one supply in a downstairs bedroom, one supply in an upstairs common area, and one supply in the master bedroom. Assume that each supply is designed to supply one quarter of the total supply necessary for the HVAC system. Now, assume that returns are equally sprinkled throughout both floors, such that there is on average two returns in each of the ten rooms in the house. Every night, mom and dad go into their 180 square foot master bedroom and shut the door. They like their privacy. And, every night, their teenage son goes into his 120 square foot downstairs bedroom and shuts the door. He also likes his privacy. They live the state of Virginia, so they have hot and humid summers. Whenever the doors are closed, the air conditioning system is drawing 50% of its supply from 12% of the house. The two supplies in the common areas are drawing the other 50% of the supply from 88% of the house. The bedrooms become negatively pressurized. The bedrooms now draw in potentially dangerous air from the attic, crawlspace, wall cavities, and subfloors to make up for the pressure difference. Additionally, the airflow from the air conditioner is reduced as the fan struggles to keep up with added friction of pulling air through the cracks and crevices of the bedrooms.

This hypothetical house is not properly designed. It is unbalanced according to our definition. If it were properly balanced, each closeable room would have its own supply sized proportionally to itself. Thus, if the house were properly designed, any given room would supply the same proportional quantity of air to the HVAC system regardless of whether or not its door was opened or closed. Still, I do not recommend undergoing an arduous and expensive renovation to fix this issue in existing homes. Rather,

I recommend finding a way to keep the doors open as much as possible.

Unfortunately, the trend towards work-from-home and learn-from-home is making the open door policy increasingly difficult for some families with unbalanced homes. There are creative solutions out there for sufficiently motivated homeowners. Anything that allows enough low-friction airflow between rooms will suffice. There are even doors on the market with privacy-focused louvers that allow for airflow while breaking line of sight.

This section will conclude with a more detailed discussion of how negative pressure is hazardous to occupant health. Throughout this book, I have recommended numerous imperfect products because, in my estimation, they appropriately balance healthfulness, accessibility, and execution risk. Core to many of my recommendations is the assumption of proper installation and usage. For example, I recommend certain fiberglass insulations despite their impurities because they are inert and encapsulated when installed appropriately. Similarly, I take no issue with gas-burning appliances because they exhaust to the exterior when installed appropriately. A similar argument holds for many materials and appliances I recommend.

The presence of negative pressure violates the assumption of proper installation and usage. Fiberglass can not be considered encapsulated under negative pressure because air is constantly running through it, picking up potentially hazardous toxins in the process, and making its way into the living space. Indoor natural gas combustion can not be considered properly exhausted in the presence of negative pressure. No matter what fan is pushing the combustion byproducts out of house, there exists another force pulling some fraction of the combustion byproducts into the house.

Negative pressure can also create or exacerbate mold growth. Significant negative pressure can pull moisture from humid outdoor air into the wall cavity where it can promote mold growth. This negative pressure can then pull spores and mycotoxins from the wall cavity into the indoor air.

8.6 Chemistry accelerators

Certain air purification devices unpredictably modify chemicals in the indoor air. I refer to these devices as *chemistry accelerators.* Every chemical in the indoor air (e.g. carpet off-gasses, fumes from household cleaning products, and outdoor pesticides) can potentially be manipulated by chemistry accelerators. In highly controlled lab conditions, these chemistry accelerators might have some beneficial effects, like the destruction of harmful microbes. But, in real indoor environments, no one is sure of how these devices interact with common indoor air pollutants.

Chemistry accelerators can break down many concerning chemicals into even more dangerous chemicals. For example, the pesticide malathion can be broken down by ozone into malaoxon. Malaoxon is much more dangerous that malathion. (Toxicity data suggests that malaoxon is hazardous at 1,000 times lower concentrations than malathion.)[6] On paper, an organic chemist could easily look at the chemical formula for a common indoor air pollutant and suggest some ways that it could react with ozone to produce new chemicals. These new chemicals are largely unknown and unstudied. But, they are very likely harmful. More than a few of them are sure to follow the pattern of hazardous decomposition exhibited by malathion.

Additionally, a large number of these indoor air pollutants can be broken down into formaldehyde by chemistry accelerators. We know this for a few reasons. First, formaldehyde is the simplest of a group of chemicals called *aldehydes.* The theory of chemistry suggests that exposing almost any chemical with an aldehyde group in it to a chemistry accelerator can result in the formation of formaldehyde as a breakdown product. Second, formaldehyde is easy to measure. Because it is so well-studied, simple, and commercially important, there exist an abundance of formaldehyde measurement technologies. These measurement devices are already in the hands of eager researchers and regulators. Third, formaldehyde is toxicologically well-studied. Almost fifty years of energetic research thoroughly establishes its sources and hazardous exposure levels. And, finally, formaldehyde is dangerous at low concentrations over long exposure periods. For this reason, regulators admonish chemistry accelerators on the basis that they produce formaldehyde.[1] Still, regulators cannot so easily acknowledge the following. Chemistry accelerators are dangerous not just because they produce formaldehyde but because they produce other unknown hazardous substances in the process of producing formaldehyde. This conclusion is a logical certainty that matters to reasonable homeowners.

Chemistry accelerators include ozone-producing ultraviolet lights, ionizers, and ozone generators. These devices can be marketed under various names like "electrostatic filters", "electronic filters", or "activated oxygen purifiers". All of these names suggest that the device aims to alter the chemistry of the air by energizing it or transforming it somehow.

Devices explicitly designed to generate ozone (also called activated oxygen purifiers) are dangerous and should be avoided at all costs. Ionizers (also called electrostatic or electronic filters) do more harm than good, partially through the production of ozone. Please note this does not pertain to polarized media filters, which do not produce airborne ions. UV-V bulbs (a type of ultraviolet bulb) are inadvisable because they produce significantly more ozone than UV-C bulbs. (UV-V and UV-C refer to the frequency spectrum of the ultraviolet radiation emitted by the bulb.) Finally, UV-C bulbs are advisable only in certain circumstances where mold exposure is of greater concern than general chemical exposure.

In general, I find that UV-C bulbs are only advisable when the home and the HVAC system have not been built with the explicit goal of eliminating the conditions for hazardous mold growth. UV-C is capable of acting as a chemistry accelerator, but the accelerant effect is suspected to be very mild. Accordingly, I suspect that the mold reduction benefits of UV-C outweigh the additional indoor air pollution risk in these cases. Specifically, I find UV-C advisable in existing homes with flex ducts. Flex ducts are difficult to clean because the spiral-shaped spine of the interior of the duct creates crevices that dust and microbes can settle into. These microbes will generally find a more permanent home to grow within the plenum or the evaporator coil. I recommend UV-C be placed over the plenum and evaporator coil for existing homes with flex ducts. I also recommend the Dynamic Air RS4 with embedded UV-C for homeowners also interested in upgrading their physical filtration media.

My theory as to the relative safety of UV-C is as follows. Ultraviolet light is capable of accelerating chemistry by three methods: (1) ozone production, (2) photocatalytic reactions, and (3) photolytic reactions. Ozone production is well-studied in UV-C, and many bulbs are certified (e.g. by UL 2998) to produce negligible ozone. If these UV-C bulbs produced significant ozone, the resultant reactions would produce (1) a lot of airborne ozone which would be easy to smell and (2) a lot of airborne formaldehyde which would be easy to measure. We tend to detect neither of these when working with high-quality UV-C bulbs. Thus, I suspect UV-C bulbs are not accelerating chemistry through ozone production.

Photocatalytic reactions are reactions that require a certain type of light (the "photo" in "photocatalytic") as a catalyst, under which two chemicals already present in the air would react. Photocatalytic oxidizers are a separate and dangerous class of air purifiers that require a specific photocatalytic substrate to function. UV-C bulbs are distinguished from photocatalytic oxidizers in that they are not designed with this chemical substrate. Thus, I do not suspect that UV-C bulbs can significantly photocatalyze chemicals in the HVAC system.

Photolysis is distinguished from photocatalysis by the absence of a catalytic substrate. Photolytic reactions are those reactions that are enabled merely by the presence of light (UV light in this case). Photolysis is the stated goal of antimicrobial UV-C. When I hope for UV-C to weaken or kill microbes, I intend for this to occur by photolysis. When I recommend UV-C bulbs, I accept the risk of chemistry acceleration by photolysis in exchange for the benefits of impairing microbial growth by photolysis. Deploying UV-C is a trade-off. As such, I avoid UV-C in new construction where we can more thoroughly control the design of the HVAC system.[5,2,3]

References

[1] California Air Resources Board. *Hazardous Ozone-Generating Air Purifiers.* URL: https://ww2.arb.ca.gov/our-work/programs/air-cleaners-ozone-products/hazardous-ozone-generating-air-purifiers.

[2] California Air Resources Board. *Hazardous Ozone-Generating Air Purifiers.* 2026. URL: https://ww2.arb.ca.gov/our-work/programs/air-cleaners-ozone-products/hazardous-ozone-generating-air-purifiers.

[3] California Air Resources Board. *Potentially Hazardous Ozone Generators Sold as Air Purifiers.* 2026. URL: https://ww2.arb.ca.gov/our-work/programs/air-cleaners-ozone-products/potentially-hazardous-ozone-generators-sold-air.

[4] Change The Air Foundation. "Mini Class Series: Understanding HVAC+D". In: (2026). URL: https://changetheairfoundation.org/understanding-hvac-d/.

[5] Douglas B. Collins and Delphine K. Farmer. "Unintended Consequences of Air Cleaning Chemistry". In: *Environmental Science and Technology* 55 (2021), 12172–12179.

[6] National Institute of Health. *Malaoxon (Compound).* 2026. URL: https://pubchem.ncbi.nlm.nih.gov/compound/15415.

Chapter 9

Electromagnetic Fields

Electromagnetic fields ("EMFs") are invisible energy fields created by the movement of electric charges. They emit from power lines, interior electrical wiring, household appliances, and wireless communication devices. The terms *EMF* and *electromagnetic radiation* (abbreviated "EMR") are frequently employed interchangeably in mitigation discussions. This chapter is solely concerned with those particular non-ionizing EMFs that are commonly encountered as a result of homesite selection and home design.

9.1 Background

The ongoing debate about EMF safety is primarily concerned with whether or not long-term exposure to EMFs at levels below regulatory safety thresholds is hazardous to human health. Few issues in environmental health have generated a more enduring and intractable controversy.

Regulatory agencies have concluded that tissue heating is the only irrefutable biological mechanism by which EMF exposure is harmful to humans. These agencies presume that any EMF exposure not strong enough to raise the temperature of human tissues is safe. Accordingly, regulators set safety thresholds solely to prevent measurable tissue heating in ordinary use. Agencies relevant to the U.S. that follow this methodology include the International

Commission on Non-Ionizing Radiation Protection (ICNIRP), the World Health Organization (WHO), the Institute of Electrical and Electronics Engineers (IEEE), and the Federal Communications Commission (FCC). These organizations, in their review of published research, find no consistent and independently replicable evidence of adverse effects on human health at EMF exposure levels below those that cause temperature elevation in human tissues.[2]

A smaller independent research community (e.g. "BioInitiative 2012") has published several thousand laboratory studies demonstrating biological effects at non-thermal exposure levels. Reported effects concern cell membranes, oxidative stress, DNA signaling, and neurology. Many of these observations have been made at intensities several orders of magnitude below current regulatory safety thresholds.[1] These phenomena are often difficult to replicate consistently across laboratories. Epidemiological follow-up studies targeting specific medical symptoms typically yield uninteresting results.

Each side of the debate has accused the other of selective citation, methodological failures, and conflicts of interest. The controversy has persisted for more than four decades and shows no prospect of resolution. No attempt will be made in this chapter to adjudicate the dispute.

However, I may assert the following with reasonable confidence based on publicly available research. First, there exists ample evidence as to the bioactivity of non-thermal EMFs. Second, the long-term human health impact of this bioactivity is poorly understood. And finally, some individuals report significant symptomatic improvements when EMF exposure is substantially reduced.

My professional experience suggests that previous severe exposure often leads to persistent individual sensitization. Additionally, the professional experience of medical doctors treating environmental illnesses suggests that sensitivity to EMFs may be a consequence of exposure to environmental toxins.[3]

Reduction of common residential exposure to EMFs is at worst harmless. Therefore, a precautionary approach is warranted in residential design, especially for households with (1) children, (2) pregnant women, (3) occupants with EMF sensitivity, (4) and occupants with prior or ongoing environmental illness.

Many EMF-aware organizations encourage homeowners to reduce their exposure to specific levels defined in esoteric units (e.g. microtesla, volts per meter, or microwatts per square meter). Effective measurement requires expensive instrumentation and effective interpretation requires deep expertise.

I believe the vast majority of homeowners (including the moderately sensitive) can prevent potentially hazardous EMF exposure through a set of simple precautionary rules expressed in terms of devices and distances. These rules derive from the combined wisdom and experience of myself, my sensitive clients, and trustworthy EMF experts.

9.2 Key Concepts

EMF exposure during sleep is more hazardous than the equivalent EMF exposure during waking. Accordingly, I will distinguish between nighttime exposure and daytime exposure in my guidance. I will use the term *nighttime* to mean *while sleeping.* I will use the term *daytime* analogously.

Low-frequency EMFs from electricity (i.e. powerlines, electrical wiring, electrical outlets, and appliances) can travel through physical barriers much more easily than higher frequency EMFs. Magnetic fields from low-frequency EMF-emitters is of equal health concern and is not affected by physical barriers. The severity of EMF exposure from low-frequency emitters will be assumed to be equal to (and part of) the severity of the corresponding magnetic field exposure in the following discussions.

Very high-frequency EMFs from telecommunications equipment (i.e. 5G devices) cannot travel easily through physical barriers. In our discussions, I will assume that anything thicker and denser than a standard brick wall can block very high-frequency EMFs. Here, I categorize very high-frequency EMFs as anything potentially above the 10 gigahertz range. As of the time of writing, this only serves to identify 5G telecommunications equipment.

Questions as to the presence of physical barriers will be ignored for the purposes of discussing the safety of everything in between the low-frequency and very-high-frequency categories. Devices utilizing this frequency spectrum are capably of communicating over very long distances and frequently exhibit pulsing behaviors. Further, these types of telecommunications devices are typically capable of modulating their signal strength to reach other far away devices. While barriers may help in reducing the hazards from this type of exposure, distance and removal will almost always override obstruction as tactics for hazard reduction.

Dirty electricity is a potentially hazardous byproduct of electrical systems that often coexists with short-range low-frequency EMFs. Ideally, electricity should flow cleanly and steadily such that its magnetic fields balance and cancel each other out. In my professional experience, simple and high-quality products that fully and smoothly use the electricity supplied to them create the least dirty electricity. Low-quality wiring or installation can spread dirty electricity throughout the home. Even if everything appears to work normally, improper grounding, lack of shielding, poor wire routing, and low-quality materials can exacerbate the hazards of existing low-frequency EMFs by producing dirty electricity. Complex modern products like dimmer switches and variable-speed motors often produce dirty electricity by unnaturally interrupting the steady electricity supplied to them.

Shielding refers to the practice of using mechanisms resembling Faraday Cages for blocking or attenuating EMFs.

Filtering refers to the practice of filtering dirty electricity from the home's electrical system using specialized devices.

GFCIs are a type of circuit breaker (ground fault circuit interrupters) that is commonly implicated in dirty electricity investigations. They are typically required by building codes in certain types of wall outlets.

9.3 Executive Summary

I recommend living at least 1,500 feet away from electrical base stations and substations. I recommend living at least 500 feet away from overhead high-voltage transmission lines (i.e. major power lines) and at least 100 feet away from all overhead powerlines. Generally, major underground powerlines are safe to walk atop of during waking hours, but I recommend sleeping at least 50 linear feet away from them. Generally, minor underground powerlines are safe to be around during waking hours, but I recommend sleeping at least 10 linear feet away from them.

I recommend maintaining a minimum of 20 feet of distance from smart meters (i.e. those that wirelessly measure electricity usage) at all times. I recommend consulting with utilities providers to replace offending smart meters with analog meters (also called "dumb meters") in the occupant's home and the homes of neighboring occupant.

I recommend using wired Ethernet in homes and avoiding all usage of WiFi. I recommend maintaining a minimum of 10 feet of distance from all WiFi-enabled devices during waking hours and 50 feet of distance during sleep. This requirement can be met by turning off the WiFi capability in specific devices or by completely turning off the devices of concern during sleep. Similarly, I recommend against all smart-home-like appliances that rely on WiFi or similar frequencies.

I strongly advice against satellite, 4G, and 5G fixed wireless access. In other words, I insist that internet access provided to the main internet modem in the house is provided over a wired rather than a wireless connection.

I recommend living at least 1,000 feet away from major cellular communications towers (i.e. those over 20 feet tall) and at least 100 feet away from minor cellular communications towers (i.e. 5G nodes). For all cellular communications towers, I recommend erecting an external barrier or situating the home such that the line of site from all parts of the home to the tower is broken. In other words, some barrier other than the house itself, like outdoor trees or another building, should obstruct the view from any part of the house to the tower. For minor cellular communications towers, I recommend a flat solid barrier at least as thick as a brick wall break the line of site from the home to the tower.

I recommend that all electrical wiring in the house is wrapped in MC cabling (i.e. metal-clad cabling) or conduit. I recommend that all plugged-in electrical appliances are situated at least 4 feet from places of common daytime occupancy and at least 10 feet from bedrooms. This distance guidance applies horizontally as well as vertically, which often complicates appliance placement in homes with crawlspaces and multiple floors.

I recommend against dirty electricity filtering except for very specialized and limiting circumstances. Further, I recommend against shielding except for very specialized and limiting circumstances.

I recommend avoiding dimmer switches entirely. For existing dimmer switches, I recommend keeping them completely turned off as much as possible, especially while sleeping. I recommend replacing any dimmer switches that provide critical lighting with regular switches.

For existing construction with romex cable shielding, I do not recommend replacement with MC cable shielding or conduit. For these existing constructions, I recommend sleeping as far as possible from wall outlets. This typically involves placing the bed in the center of the room. If this is impractical, I recommend orienting beds such that the sleeper's head rests at least two feet away from any electrical outlet.

I recommend completely avoiding solar power (namely panels and inverters) on, in, and around the home. I recommend completely avoiding electric car charging ports.

I recommend replacing microwaves and electric ranges. They are both high emitters, and their usage requires such close contact over such a long time that there is no particularly safe way to use them.

For sensitive individuals living in existing homes where following this guidance is intractable, I recommend relocation.

I do not recommend the following commonly suggested EMF safety activities. I do not recommend replacing variable-speed appliances with single-speed appliances. I do not recommend replacing GFCIs. I do not recommend replacing LED bulbs with low-tech alternatives. I do not recommend disabling electricity to bedrooms at night. I do not recommend disabling electricity to specific parts of the house at the breaker panel. In my experience, all of these activities have a low return on investment for homes wired with romex. Additionally, these activities have negligible measurable benefit for homes wired with MC cable or conduit.

9.4 Shielding

EMF shielding involves using specialized fabrics and paints to diffuse EMFs. I recommend avoiding them. Shielding poses a reflection and amplification risk.

EMF shields work like the see-through doors in microwaves. In a shielding fabric, finer weaves can block shorter waves. Any waves longer than the gaps in the weave theoretically cannot penetrate the shield. Just like in a microwave, this risks reflecting EMF sources that come from within the shielded area back towards the source. In today's connected world, anyone can carry a high-powered EMF emitter in their pocket. Additionally, smart home appliances are so ubiquitous that even experts will fail to identify some. Accordingly, the risk of accidentally magnifying occupant exposure through the use of shielding is too great to justify its use.

Interestingly, shielding cannot block low-frequency EMFs that emit from electrical systems. Given that it is easy to remove WiFi and cellular devices from a bedroom, but hard to remove electricity from a bedroom, shielding rarely provides any benefit.

EMF shielding is most justifiable in severe and densely populated circumstances where an uncontrollable external source cannot be disabled by the occupant. For example, if a telecommunications company installed a 5G tower outside the window of an apartment building, the occupant could justifiably place a lot of shielding on that window. The same might go for a neighbor in an apartment building who decides to install a 5G fixed wireless access point opposite the occupant's bedroom wall. In both of these examples, I advise against creating a cage-like shield that attempts to block all walls of a room. Rather, I recommend using shielding as a barrier against the offending uncontrolled emitter.

Densely packed neighborhoods may produce similar quagmires. I recommend coordinating with neighbors where possible to address these concerns. Accordingly, shielding is almost always inadvisable for a single-family home, even in a densely packed neighborhood.

9.5 Filtering

Filtering is a bit of an esoteric concept in EMF safety. Some products claim to be able to filter dirty electricity with a device that can be plugged into an electrical outlet or installed at the breaker panel. In my experience and testing, these devices operate less like filters and more like converters. In the process of attempting to filter dirty electricity, they create a magnetic field that runs along the neutral line of the wiring system. In effect, these devices convert dirty electricity to magnetic fields. This may be desirable in some highly specialized circumstances with highly specific sensitivities, but I generally do not recommend them. I instead recommend that the root cause of the dirty electricity is addressed by other means. And, I recommend relocation if our other recommendations are insufficient.

9.6 Smart Meters

Smart meters are electricity meters that beam data about the home's electricity usage to a centralized server supposedly for the purposes of utility billing. I say "supposedly" here because repeated field testing indicates that smart meters actually send much more data than necessary back to servers owned by utility companies. For the purposes of utility billing, only one number per month is needed (the home's electrical usage measured in kilowatt-hours). For some reason, smart meters beam detailed data about the home's electrical usage every 10 or 15 seconds like clockwork. EMF metering demonstrates that these pulses are spiky and powerful, seemingly carrying a lot of data a pretty far distance. These meters typically communicate on a frequency range a little below the standard WiFi spectrum (around 1.5 gigahertz), so they have a health impact comparable to a very broken and erratic wireless router.

The shape, strength, and frequency of the signal emitted by smart meters makes them much more dangerous than other devices that emit on the WiFi-like spectrum. Spiky emissions are more biologically disruptive than smooth emissions. Long-distance emissions are necessarily more powerful and hazardous, as well.

Unfortunately, the transition from analog meters to smart meters has resulted in some very hazardous arrangements of smart meters in the wild. Before the advent of smart meters, an electrical utility employee would visually inspect the kilowatt-hours number on the meter once per month to calculate the utility bill. It made logistical sense for all of the analog meters to be clustered together in a single room for easy inspection. When the analog meters were swapped out for smart meters, the smart meters stayed in the same rooms and inherited the same densely clustered layout. This is why we often see large clusters of dozens of smart meters in utility closets and alleyways of office building and apartments. These meter banks are extremely dangerous because each meter emits a separate signal to the server independently. Smart meter banks have injured a lot of people. Many smart meter banks are on walls shared by bedrooms and office spaces. As the story often goes, the utility company upgrades the meters one day, and then the symptoms start. At least a few people have devoted their lives to EMF safety consulting as a result of this sort of experience.

Fortunately, single-family homeowners have options. In my experience, a simple call to the utility company will allow single-family homeowners to swap their smart meters for analog meters at their own expense. Homeowners should expect to be charged one-time swap fee of up to a few hundred dollars and monthly meter-reading fee that can range from 10 to 50 dollars per month.

9.7 GFCIs

Some EMF experts claim that GFCIs are a significant source of dirty electricity. Based on my experience and general understanding of dirty electricity, they are probably right. GFCIs resemble other complex electrical devices that manipulate the electrical current and produce dirty electricity. GFCIs are sophisticated breakers that constantly monitor the flow of electricity into an outlet. They are designed to trip and break the flow of current if they detect conditions that would lead to electrocution of the user.

GFCIs are required in many types of outlets by building codes. Some electricians install them in all outlets out of an abundance of caution because they are cheap and effective.

Some EMF experts will instruct occupants to rip out GFCIs after the home has passed inspection. I advise against this. A licensed electrician would not do this on the homeowner's behalf because it violates building codes, so the homeowner would have to do it himself. This leaves the homeowner in the precarious position of modifying his electrical system to remove code-required safety features. I believe this operation poses substantial risk with only marginal potential benefit, so I advise against it. Large-scale upgrades to electrical wiring systems like advanced cable shielding are much more impactful than the removal of GFCIs. Accordingly, if a comfortable environment cannot be achieved with the other recommendations set forth in this chapter, I recommend relocation.

9.8 Variable speed motors

Some EMF experts have demonstrated that variable speed motors in appliances produce a lot of dirty electricity. I am inclined to believe them. These devices manipulate the electrical current just like every other device that produces dirty electricity. Many mechanical appliances we use in our homes can have variable speed motors. Examples include blowers in HVAC systems and agitators in washing machines.

Proponents of single speed motors also assert that variable speed motors are a pointless luxury. For example, a clothes dryer can be built with a single speed motor. Such a clothes dryer would only spin at one speed. Such a dryer is probably just as useful as a variable speed dryer, but it is also safer. The same argument goes for virtually every other appliance in a house.

In my experience, there exists no practical way to determine whether or not an appliance contains a single speed motor. Manufacturers and retailers often do not know, do not care, or have no way of knowing which type of motor a product contains. Further, the modern supply chain is replete with variable speed motors. Variable speed motors are seen by the industry as a strictly superior alternative to single speed motors. As a consequence, motor manufacturing has mostly shifted towards variable speed, and appliance fabricators do not have much of a choice as to which type of motor they source for their products. This problem is simply not within the power of the average homeowner to solve. Accordingly, we recommend managing dirty electricity exposure by maintaining personal distance from large appliances. I also recommend homeowners invest time and effort into other EMF-safe improvements to their homes.

9.9 Fixed Wireless Access

Some internet service providers ("ISPs") have begun to solve difficult routing problems by offering fixed wireless access ("FWA") points. Avoiding these is simple but worth discussing. An FWA connection is simply a modem that interacts with the rest of the internet over some wireless communication protocol (satellite, 4G, or 5G) rather than a traditional underground cable. If an ISP cannot reach a location via underground cables, it may try to reach that location via FWA. As a reasonable homeowner, simply insist on connecting to the internet via underground cables rather than FWA. The power of EMFs from an FWA is likely in between that of a large WiFi router and a small 5G node. An FWA would necessarily have to be placed on the homeowner's property, so I do not see a safe way to utilize them.

9.10 Cellular towers and nodes

There are two classes of outdoor cellular communications devices that reasonable homeowners need to be aware of. Large cell towers (called "base stations") contains between 8 and 32 or more large canister-shaped nodes arranged in a pattern. These large towers are typically between 20 and 100 feet off the ground. They can be found on free-standing poles, often near highways. They can also commonly be found on top of hotels, churches, and schools.

In my experience, it is not worth speculating about what type of EMF fields are emitted by these large devices. They are almost certainly 4G, 4G LTE, and 5G towers. They may also provide support to normal phone calls and text messages that may or may not operate on the same or older frequency spectrum. Regardless, the towers are designed to project wireless access for long distances at high power to thousands of users. Accordingly, significant distance is warranted. We recommend at least 1,000 feet of distance between the house and the base station. Additionally,

I recommend the line of site to the tower is broken by something other than the home itself (e.g. other buildings, external barriers, or trees).

The second class of outdoor cellular communication device is the 5G node. 5G nodes are small single canisters about 3 feet tall and 6 inches in diameter that are often found atop light posts. They will generally be between 10 and 20 feet off the ground, but they can be higher up in densely populated areas. Again, it is not worth speculating exactly which types of frequencies are emitted by these devices. I can be reasonably certain the most health-relevant function of these towers is 5G telecommunication. Many 5G deployments use WiFi-like frequencies (around 2.5-6 GHz) and some use high-band millimeter waves (24-71 GHz). (Typical 4G bands operate below 3 GHz). Higher frequencies enable faster data transfer but are also more vulnerable to distance and obstruction. This higher frequency warrants higher overall precaution due to limited long-term research on these bands. Accordingly, I recommend a minimum of 100 feet of distance. Additionally, I recommend breaking the line of sight with something at least as thick as a brick wall (e.g. other buildings or external barriers).

I recommend boots-on-the-ground analysis over web-based tools that map telecommunications equipment. Typically, these web-based tools pull from publicly accessible registration databases that are not necessarily comprehensive. Web-based tools typically only surface a subset of all telecommunications equipment. They often miss important small nodes that should factor into homesite selection.

9.11 Powerlines

Reasonable homeowners need to be aware of the various types of power lines and methods for classifying them. Each type warrants a different form of precaution. Large power lines (called "high-voltage transmission lines" or "HVTLs") typically tower 20 to 50 feet off the ground. They typically have at least three large cables running in parallel, often more. There is a specific reason these towers are so high and so broad. Understanding the reason helps us identify the level of precaution warranted for different types of powerlines.

One of the greatest operational hazards of HVTLs is electrical arcing. Electrical arcing describes the temporary short-circuiting of electrical equipment by arcs of electricity that jump across the air from cable to cable. The electrical system is a big circuit, and to get the circuit to terminate at the proper place (the home), the entire network of power lines running up to it needs to not short-circuit (or arc). The electrical wires used in homes and offices are typically insulated with something, be it rubber sheathing or otherwise, that allows them to run together without arcing or short-circuiting. In HVTLs, the insulation is the air. This is the reason simple lamp cords are insulated with rubber but the big electrical cables overhead are not.

HVTLs are intentionally designed to use the ambient air as insulation. Electrical engineers carefully size the air gaps between the cables to achieve this effect. This is the primary reason HVTLs are so tall and broad. Economically, this makes perfect sense. Adding air gaps is cheaper than adding insulative coatings.

For us as reasonable homeowners, the air gaps are the key to understanding the hazards posed by EMFs from power lines. Across the country, there are a few different designs of HVTLs. In the mid-Atlantic, they are often arranged on large hexagonal frames. In Texas, they are often stacked vertically on a monolithic pole. Both types of power lines pose essentially the same hazard. Both can be identified by the large distance between the cables (approximately 15 feet).

The size of the air gaps is typically proportional to the size of the magnetic field emitted by the cables. Field tests suggest that, all else being equal, the health-relevant magnetic field grows larger as the air gap between cables increases. HVTLs are the most extreme example of this phenomenon in the wild. Accordingly, I recommend living at least 500 feet away from them.

Next, I will discuss overhead distribution lines. Overhead distribution lines are smaller than HVTLs. They are designed to serve a smaller number of households. Overhead distribution lines can vary wildly in size and hazard. Out of precaution, I recommend living at least 100 feet away from them. The worst overhead distribution lines I have ever seen were uninsulated and air-gapped at about 4 feet. They produced potentially hazardous magnetic fields in a 100-foot radius. My very cautious guidance arises from this experience.

Less precaution may be necessary for smaller distribution lines. I recommend assessing distribution lines according to their air gap size, the presence of insulation, and the size of the community they serve. Smaller air gaps, insulated lines, and smaller communities typically translate into smaller magnetic fields. Less distance is justifiable in these scenarios. For example, many houses have a single insulated overhead line running from their house to the electrical grid. These are likely of little concern.

Finally, I will discuss underground lines. In general, I love underground power lines. I highly recommend that homeowners bury their powerlines if at all possible. Buried powerlines have thick multi-layer insulation. In my experience, the industry standard insulation for buried powerlines almost completely neutralizes the magnetic fields.

Underground powerlines, in their default configuration in the wild, will rarely ever require corrective action. I believe underground powerlines are safe to be around during the daytime but require some distance during the nighttime. Almost all major underground powerlines will already be far enough away from most sleeping areas (at least 50 feet). The same is true for minor underground powerlines (at least 10 feet). Generally, underground powerlines may be hard to locate. I classify any major underground powerlines as those that substitute for HVTLs, and I classify minor underground powerlines as everything else. Accordingly, a major underground powerline can be identified by a large tract of linear deforested path with signage around it indicating the presence of powerlines. Minor underground powerlines can be identified by the presence of houses and the absence of overhead powerlines (i.e. the power has to come from somewhere). For prospective homeowners shopping in suburban areas, the presence of overhead distribution lines can be an easy way to disqualify a neighborhood.

Finally, avoid living near electrical base stations and substations. I recommend at least 1,500 feet of distance. This recommendation effectively disqualifies homeowners from living in the same neighborhood as a base station or a substation. The stations themselves emit large magnetic fields. And, each station has a multitude of HVTLs coming in and out of it. These fields compound on each other to create a very large radius of potentially hazardous magnetic fields. My recommendation is very precautionary. The magnetic fields around base stations and substations vary dramatically in the wild.

9.12 Solar

I recommend against single-home solar power systems. From an EMF safety perspective, a single-home solar power system is like a miniature electrical substation. The solar panels, inverters, and batteries all emit EMFs. Common household solar power systems use the roof for the panels and the utility closet for the inverter. Any individual component of the solar power system can be an EMF hazard if it is located close to a sleeping area.

A sophisticated homeowner may be able to build a safe solar power system by separating the components from the house. This would likely require a large plot of land and some careful engineering.

References

[1] BioInitiative Working Group. *BioInitiative Report 2012: A Rationale for a Biologically-based Public Exposure Standard for Electromagnetic Fields (ELF and RF)*. Ed. by Cindy Sage and David O. Carpenter. Updated research summaries available as of 2022; self-published online report by an international group of scientists and public health experts. 2012. URL: https://bioinitiative.org/ (visited on 12/23/2025).

[2] IEEE. *IEEE Standard for Safety Levels with Respect to Human Exposure to Electric, Magnetic, and Electromagnetic Fields, 0 Hz to 300 GHz*. Approved 4 October 2019 by the IEEE Standards Association. New York, NY, USA: Institute of Electrical and Electronics Engineers, 2019. DOI: 10.1109/IEEESTD.2019.8859679. URL: https://ieeexplore.ieee.org/document/8859679.

[3] Neil Nathan. *Toxic: Heal Your Body from Mold Toxicity, Lyme Disease, Multiple Chemical Sensitivities, and Chronic Environmental Illness*. 2nd. Paperback. Victory Belt Publishing, 2025. ISBN: 978-1-62860-566-2.

Chapter 10

Water and Appliances

This chapter will address some controversial issues related to water and appliances that are not addressed elsewhere in the book.

10.1 Natural Gas Ranges

There exists an ongoing debate about the safety of gas stoves and ovens relative to their electric counterparts. Critics of gas stoves and ovens assert that the indoor combustion of natural gas contributes significantly to indoor air pollution. I will explain my stance on the controversy. I recommend properly ventilated gas ranges over electric ranges. Here, I use the term *range* to refer to a single appliance that combines a stovetop and an oven into one unit that uses the same heat source.

In the previous chapter, I recommended that homeowners keep actively used electrical appliances at least 4 feet away from places of common occupancy to protect themselves from potentially harmful EMF exposure. This guidance is easily applied to large appliances that homeowners do not spend much time directly interfacing with. These might include air conditioners, ERVs, and dishwashers. I distinguish these passive-use appliances from active-use appliances. Active-use appliances are those that homeowners spend a lot of time up close and personal with. Active-use appliances include things like stovetops and hair dryers.

The EMF risks of active-use appliances cannot be mitigated by distance, because their usage requires close proximity to the operator. In other words, a chef sauteing vegetables on a stovetop will be up close and personal with that saute pan for a significant amount of time. The associated EMF exposure cannot be reasonably mitigated.

Unventilated indoor combustion of natural gas certainly contributes to indoor air pollution. For example, researchers have found that benzene (a very hazardous aromatic hydrocarbon and VOC) can be a byproduct of indoor natural gas combustion.[2] Fortunately, this pollution risk can be mitigated by proper ventilation.

Interestingly, the majority of indoor air pollution generated from cooking with natural gas is attributable to cooking itself — not natural gas combustion.[3] In particular, fats, oils, vinegars, and fibrous compounds break down during cooking and emit a large variety of indoor air pollutants.[1] Swapping from a gas range to an electric range only mitigates a some of total indoor air pollution generated from cooking. In either case, sufficient continuous ventilation is required to avoid indoor air pollution during cooking. Accordingly, I recommend thoroughly ventilated gas ranges over electric ranges.

Thorough ventilation is best achieved through a range hood. A range hood is a relatively inexpensive appliance that sits about three or four feet above the range and continually exhausts air from the interior to the exterior. Proper usage of the range hood is critical in order to realize its benefits. The range hood should be turned on before the first burner is turned on. It should be turned off after (1) the last burner is turned off and (2) the cookware cooled down enough to avoid further vaporization of cooking residues.

It is worth mentioning that range hoods are situated close the head of the cook and are themselves an EMF-emitting electrical appliance. While this is true, range hoods are relatively low-power appliances when compared to electric ranges. Accordingly, they emit weaker EMFs. Ultimately, we must balance the risks inherent to cooking to achieve an optimal outcome given that some EMF exposure and indoor air pollution is inevitable.

Given this information, I trust that homeowners can adjust their kitchens in accordance with their comfort levels with various types of exposure. Sensitive individuals armed with this information are encouraged to adapt their kitchens and cooking styles to meet their personal needs.

Hypothetically, a homeowner with a strong urge to minimize inflammogenic exposures during cooking might find comfort in some uncommon cooking practices. For example, grilling outside avoids indoor natural gas combustion. As another example, passively boiling a big pot of stew on an electric range avoids most EMF exposure. An educated and sensitive homeowner can adapt his behavior to achieve safe cooking practices.

10.2 Natural Gas Fireplaces

See the previous section for a discussion of the risks of indoor natural gas combustion. The same safety precautions apply to natural gas fireplaces.

Natural gas fireplaces can be safe and harmless appliances if installed and maintained properly. Natural gas fireplaces needs to be fully sealed such that no detectable natural gas or combustion byproducts enter the living space. Homeowners using natural gas fireplaces need to perform a gas sniffer check annually on the natural gas plumbing. Homeowners should use sensitive carbon monoxide meters, VOC meters, and gas sniffers to ensure that no natural gas is leaking from the fireplace, even when it is not in use. Homeowners that do not wish to inherit the testing and maintenance burden of existing gas fireplaces can simply turn them off. Additionally, always follow all applicable laws and building codes regarding installation and operation.

10.3 City water processing

The water system of a house can run on either city water or well water. I will discuss sensible default water processing strategies for city water.

For houses on city water, I recommend whole-home filtration and softening. I recommend that homeowners use whichever systems are familiar to their choice of trustworthy plumber. Any system with a significant amount of carbon filtration and softening capacity will be sufficient, provided the filter capacity matches with the inflammogens found in the water report. I do not have strong product-specific preferences for these systems.

I do not recommend ultraviolet purification systems for homes on city water. The risk of microbial contamination in city water is fairly slim. Rather, the main health risks of city water are the chemicals added to it by the city.

I do not recommend reverse osmosis systems for homes on city water. It is fairly common for homeowners to install reverse osmosis systems at the point of service (e.g. at the refridgerator water dispenser or the kitchen sink). These homeowners tend to believe that their drinking water needs a higher level of filtration that only reverse osmosis can provide. This is somewhat true, but reverse osmosis is not without its downsides. Reverse osmosis filters work very well at purifying water. There are certain toxins commonly found in city water that can only be removed by homeowners with reverse osmosis. Unfortunately, reverse osmosis water filters also strip all beneficial minerals from the water. It purifies the water a little too much. Drinking highly purified water can leach minerals from the body, leaving it starved for electrolytes.

Remineralization systems attempt to address this problem. Remineralization systems add minerals back to the water at the point of service (after it has undergone reverse osmosis). I have been underwhelmed with the performance, consistency, and cost of these remineralization systems. The remineralization cartridges tend to deposit inconsistent quantities of the minerals into the water. And, they create a constant recurring expense, as the refills must be purchased and installed regularly.

Instead, I recommend that homeowners on city water systems who require purity, purchase high quality spring water for all of their drinking and cooking requirements. This practice costs about $2,000 per year per member of the household, depending on the brand. My reasoning is as follows.

Health-conscious homeowners already spend a lot of money on food. Depending on the region, the cost of healthfully feeding a family of four can easily reach $30,000 to $40,000 per year. For such homeowners, the best indicator of the quality of food they buy is its proximity to nature. Less processing is always better. I believe homeowners should treat drinking water the same way.

Many health-conscious homeowners on city water will use a maxed out reverse osmosis setup. In such a setup, their drinking water undergoes (1) carbon filtration, (2) softening, (3) reverse osmosis, and (4) remineralization. All of that happens after the water leaves the city plumbing. All of that attempts to undo the side effects of dozens of mysterious processing steps performed by the city. That remineralized reverse osmosis water might be the best anyone can do with city water, but it is still highly processed Unforunately, no one knows of all of the potentially harmful chemicals in city water, so no one can test to see if reverse osmosis removes all of them. A similar principle holds for remineralization. The water undergoes a lot of unnatural transformations, so the potential for unknown negative effects is high. While other benefits of spring water exist, I believe this reason alone justifies the increased expense of drinking it exclusively.

Non-drinking water requires merely carbon filtration and softening. An enumeration of all of the hazards of city water is well beyond the scope of this book. However, one specific hazard is worth mentioning — chlorine and its dermal toxicity. Almost all city water contains some amount of chlorine. Bathing in chlorinated city water is more acutely hazardous than drinking it. Fortunately, common whole-house carbon filtration and softening will completely address the dermal toxicity risk of chlorine.

First, the carbon filtration almost completely removes the chlorine. Then, the softener reduces the oil-stripping ability of the water. We all have a natural oil-based protective barrier on our skin. When we bathe, some fraction of this barrier is lost. If we lose too much of it, our raw skin becomes exposed to toxin penetration from the water itself. Soft water is much gentler on this barrier. With this barrier maintained, any remaining toxins in the filtered water have a harder time penetrating our skin. In combination, softening and filtration almost completely prevent toxins in city water from penetrating our skin. For most people, this approach will eliminate symptoms of dermal toxicity caused by bathing in city water.

10.4 Well water processing

I believe it is difficult to give standardized advice for wells. A deep well can be an amazing water source. I recommend consulting with local experts to test likely contaminants. Many wells require basic filtration to address unpleasant groundwater minerals such as iron and sulfur. Unfortunately, some exotic contaminants in well water can only be addressed by proper site selection. Carbon filtration, softening, reverse osmosis, and remineralization are all moot issues for deep wells. Whole-system ultraviolet light is almost always advisable to protect against the emergence of coliform bacteria. While I confidently recommend well water systems, each system is unique. Accordingly, I cannot give much specific advice here.

References

[1] ARCADIS Geraghty & Miller, Inc. *Indoor Air Quality: Residential Cooking Exposures.* Tech. rep. Prepared for California Environmental Protection Agency, Air Resources Board. Sacramento, CA: California Air Resources Board. URL: https://ww2.arb.ca.gov/sites/default/files/2020-03/cookingstudy_0.pdf.

[2] Yannai S. Kashtan et al. "Gas and Propane Combustion from Stoves Emits Benzene and Increases Indoor Air Pollution". In: *Environmental Science & Technology* 57.26 (2023). Stanford-led study detecting benzene in natural gas combustion byproducts, pp. 9653–9663. DOI: 10.1021/acs.est.2c09289.

[3] Daniel Tormey and Steve Huntley. *The Effects of Cooking on Residential Indoor Air Quality: A Critical Review of the Literature with an Emphasis on the Use of Natural Gas Appliances.* Tech. rep. Available at https://cbia.org/wp-content/uploads/2023/03/ANALYSIS-Effects-of-Cooking-on-Indoor-Air-Quality-3.2.2023.pdf. Connecticut Business & Industry Association (CBIA), 2023. URL: https://cbia.org/wp-content/uploads/2023/03/ANALYSIS-Effects-of-Cooking-on-Indoor-Air-Quality-3.2.2023.pdf.

Chapter 11

Mental Health

I regret to burden my readers with this knowledge. We as homeowners cannot eliminate all exposures to all toxins. We must learn to be at peace with some exposure.

I would like to extend a very special thanks to Dr. Morgan Cowger, for providing valuable insights in support of this chapter.

Dr. Cowger is a board certified Psychiatrist who sees adults of all ages. She attended medical school at Western University of Health Sciences, College of Osteopathic Medicine of the Pacific-Northwest, then completed her general Psychiatry residency at the University of Hawaii. She feels passionate about identifying underlying factors that contribute to psychiatric symptoms, and practices functional Psychiatry after completing a fellowship through Psychiatry Redefined.

11.1 Stress as an Inflammogen

Information like this can be so overwhelming that stress ends up causing the reader more inflammation than environmental toxins. There are many encyclopedic books about environmental pollutants on Amazon. There are thousands and thousands of chemical-specific research papers on PubMed. There are dozens of prohibitively expensive alternatives to every common building material. A healthy home adventurer trying to fit all of the puzzle pieces together often gives up in frustration.

After some research, he realizes, "The only safe house is a 3 million dollar stone castle in the countryside." When he greets the other extreme, he thinks, "I should just live in a tent in the woods until I can figure all of this out." He ultimately concludes, "I should just build a cinderblock box in my backyard."

We need to be realistic in the face of difficult circumstances. That castle is not only extremely expensive, it is also impossible to build using the current supply chain. That camping excursion is not just inconvenient, it also renders you homeless. That cinderblock box might seem cheap and simple, but it is also illegal to inhabit. All of these solutions are technically possible. (A concerning number of victims of environmental toxicity try the cinderblock box approach.) Yet, they are all extremely stressful. Stress is its own source of inflammation, and no one can heal while under significant stress.

Any publication that recommends impossible solutions to addressable problems is alarmist noise. I encourage readers to only consume relevant and actionable information about healthy home design. Actionable information is the type of information that gives the homeowner just enough information to make an informed decision that meaningfully improves the environmental health of his home. Non-actionable information directs the reader's attention to unachievable solutions. Non-actionable information lives on the border of the aspirational and the theoretical. Non-actionable information clutters the reader's mind with useless technical details. This sort of information should be discussed in the halls of progressive architectural colleges, not espoused as the everyman's solution to healthful living.

For example, while formaldehyde is dangerous, it is also natural and ubiquitous. On one level, demanding a formaldehyde-free home would be demanding a wood-free home. On another level, banning all formaldehyde-based resins in a new construction triples the total construction cost. Unqualified formaldehyde alarmism often creates more inflammation through stress and anxiety than it solves through positive materials choices. A homeowner that believes her phenol-formaldehyde sheathing and subfloor is harming her will never be relaxed enough to heal from anything.

We have specifically seen examples of healthy home influencers advising against the use of softwoods (like pine and cedar) because these influencers consider terpene (the chemical responsible for cedar's pleasant smell) to be a VOC of comparable hazard to formaldehyde. This is insane. While terpenes are a VOC, natural terpenes are harmless (and often therapeutic) when released from fresh softwoods. Virtually all structural wood (including framing lumber and sheathing) is made of pine containing naturally occurring terpenes. These influencers advise readers to avoid softwood furniture to minimize their exposure to terpenes despite the fact that the the reader's entire house is likely framed in pine OSB and 2x4s. As reasonable homeowners, we simply do not need to push the envelope on every single issue all at once — especially when many so-called issues are toxicologically irrelevant.

I advise against purchasing handheld VOC meters for similar reasons. I have already talked at length about VOC alarmism and the false premise of low-VOC marketing. Generally, unless approached with a deeply sober mind, handheld VOC meters produce more anxiety than insight. All of the VOCs in the universe are innumerable. Airborne VOCs in the home can come from natural gas combustion, the vaporization of cooking oil, evaporation of harmless solvents from cosmetics, the passive storage of alcohols and fuels, and the pleasant smells of natural softwood. Rarely if ever will a VOC meter allow a homeowner to detect the potentially harmful VOC molecules leaping off of the dried paint on the walls. (In any case, I do not believe VOCs are the primary toxin of concern in paint.) A homeowner that chases perfection on a generic VOC meter might well be driven mad. It bears repeating that no one can heal in a state of extreme stress.

Further, I detest scare tactics based on scary-sounding latinized chemical names. All too often, healthy home advice sounds a little too much like the following.

> Dihydrogen monoxide (DHMO) ... is a chemical compound that is commonly used in lab investigations. This chemical compound, sometimes referred to as hydric acid, is colorless and odorless. DHMO is known to be a part of many environmental hazards, toxic substances, diseases, and disease-causing

> agents. Each year DHMO is listed as a contributing factor in thousands of deaths, and it contributes to millions of dollars of damage to personal property and the environment. However, DHMO continues to be used widely by industry, government, and consumers worldwide. See [below] ... for a summary of the uses, effects, and hazards of DHMO.[1]

The above example might seem like just a cute joke. Most readers would be unperturbed if we told them that the air they breathe was full of dihydrogen monoxide. Unfortunately, we are all vulnerable to this type of information when our minds and bodies are already in a state of high alert. This type of information can cause us panic, stress, and anxiety.

The technical information in the above passage is factually accurate. The tone is cold and direct. Somehow still, the delivery is scary. The juxtaposition of facts paints a fearsome picture in our minds. In more words than necessary, the passage tells our brains that dihydrogen monoxide is (1) esoteric, (2) dangerous, (3) ubiquitous, and (4) unavoidable. How does a reasonable person respond to such information? Obviously, the rational mind seeks to control the only factor it can. No one can make the threatening chemical less dangerous, less ubiquitous, or less unavoidable. But the rational mind can attempt to remove the esoterism of the threatening chemical by seeking more information about it — by studying it. In our current day and age, dense and low-quality technical information is abundant. Unfortunately, so are threatening chemicals. It is no wonder homeowners are driven mad in the quest for knowledge.

Homeowners often shoulder the burden of knowing everything in order to protect themselves and their families from everything. They see each toxin as an invisible predator. They study the predators in hopes of outsmarting them. But, there are uncountable invisible predators. Every day, these homeowners rise to meet this unconquerable objective where the stakes are life-or-death. Living like this, never giving the body a moment to rest, can break a person down.

11.2 Treating one problem at a time

Leading medical practitioners treating environmental toxicity have made the following important observation. Attempting to treat anything other than the primary cause of an illness will fail to heal the patient.[2] I impute this principle to healthy home design. Homeowners are encouraged to treat one problem at a time. Each action taken to improve the health of an indoor environment should attempt to treat the most significant outstanding hazard in that environment. Influencer-grade health alarmism defies this principle. It is difficult to imagine, for example, that pine terpenes from a solid wood table will ever be the number one most relevant health hazard in a home. Accordingly, a homeowner that has substantially complied with the guidance in this book would be well-advised to also seek non-environmental explanations to their problems. As each environmental concern is sequentially mitigated, additional environmental explanations become increasingly implausible. And, as mitigation projects become increasingly desperate, so does the likelihood of inflicting unnecessary stress on oneself.

The process of determining the primary cause of an illness is complex. It requires a lot of wisdom and judgement from specialized medical professionals and healthy building experts. Unfortunately, diagnosing certain illnesses requires a lot of awareness and initiative from the patient. Consider, for example, some of the most common root causes of generalized autoimmunity (e.g. mold toxicity, heavy metals toxicity, and tick borne illness.) The total cost of testing for these conditions (averaging in all medical consultations and lab fees) is over $1,000 each as of the time of writing. Therefore, no patient is likely to be tested for these conditions as part of a routine checkup. Generally, a patient will only ever be tested for these conditions by a specialized physician after receiving a referral from a non-specialized physician. By the time the patient gets the referral, the patient has likely already spent years unsuccessfully treating the illness.

For example, if the patient is (1) experiencing autoimmune symptoms, (2) aware of a source of personal mold exposure, and (3) aware of the deleterious health effects of mold exposure, only then will the patient discuss mold exposure as a risk factor with his medical provider. At this point, the medical provider will likely take steps to get the patient tested. This hypothetical scenario is the knowledgeable and empowered patient's story of a successful diagnosis with mold toxicity. In my experience, this is an uncommon story. The following is a more common story.

Typically, a patient will have confusing autoimmune symptoms for some time. She will try a lot of common tests and treatments that will ultimately fail. Then, she will find herself drawn to alternative medical providers with expertise in autoimmunity. During this process, the patient's condition has worsened. The patient might feel like she will fall apart if she can not identify a root cause and make meaningful progress in her health immediately. Next, the patient starts to develop a relationship with an autoimmune health expert that is capable of testing and treating a multitude of possible root causes. The patient is presented with roughly the following bargain. The patient can spend over $1,000 to attempt to identify one possible cause at a time (e.g. mold toxicity) or $10,000 to test many possible causes at once (including tick borne illness and heavy metal toxicity and many others). The most critically sensitive patients will opt for the most expensive and comprehensive approach. Therein lies the economic tragedy of the patient's journey.

The patient's reasoning is simple. Once the patient reaches that critically sensitive point in her health, when her autoimmunity becomes a life-or-death issue, the possibility of failing to detect a single root cause becomes unbearable. So, the patient spends a great deal of time and money testing absolutely everything. Unfortunately, this story is very common.

An alternative economic tragedy is also common in patients with Multiple Chemical Sensitivity ("MCS"). MCS is a condition where nearly all chemical odors, even toxicologically benign odors, can trigger symptoms of general unwellness (e.g. nausea, dizziness, headaches, fatigue). The smell of cork is a an example of a toxicologically benign odor that commonly triggers symptoms in MCS patients. MCS is many things, but it is not a root cause. Rather, it is a cluster of symptoms. Of course, living with MCS can be very frustrating.

The economic tragedy of MCS results from a radical attempt to eliminate every possible trigger. For example, MCS patients are well advised to replace all of their irritating household cleaners with natural alternatives that do not trigger their symptoms. Conversely, if an MCS patient is fortunate enough to have natural cork flooring, that patient would be well advised to try to treat the root cause of his MCS rather than replace his flooring. At some point, the marginal benefit of replacing triggers becomes negative.

MCS or otherwise, we are all prone to falling into seek-and-destroy loops with our homes. We learn about a problematic toxin (to "seek"), then we make a healthy substitution (to "destroy"). Whenever we are evaluating a healthy substitution, we have to first ask ourselves some hard questions. First, "Am I confident that this thing I am about to do is the highest-value activity available to promote the health of myself and my family?" Then, "What health outcome can I monitor to inform me that I have improved my home enough?" If we as homeowners do not know what our objectives are, then we do not know when to stop improving our homes. If we never stop improving our homes, we will eventually run out of time and money. The home is merely one arena of one battle (the health battle). We need our time and money to fight other battles in other arenas, too. We are well advised to treat one problem, the most prominent problem across all arenas, at a time.

11.3 Feeling safe and loving your home

I invite readers into following thought experiment. Imagine you are in your home. Imagine you feel safe. Imagine feeling that your home is beautiful. Imagine being proud of your home. Imagine feeling proud of yourself for placing your loved ones in that home. What does that home look like to you? What shape is it? What colors are the floors and walls? What kind of furniture do you have? Where do you spend the most time in that home?

Ugliness and destitution are not stops on the road to healing. In my experience, homeowners have to feel safe and love their homes in order to heal. They do not just have to be safe. They also have to feel safe. Sometimes, feeling safe is a journey of discovery and education. Sometimes, highly analytic types only feel safe if meters say so. Everyone is different, but no one is hopeless.

If a homeowner's carpet is particularly toxic, the objectively safe thing to do is to pick it up and rip it out right now. That act is within most anyone's physical abilities, and it will immediately improve the indoor air quality. However, OSB subfloor is really ugly. The home will look like a construction site until the homeowner can spare the time and capital to do something about it. Unless the homeowner's health woes are particularly acute and particularly related to carpets, staring at OSB subfloor for months might be too ugly to bear. I implore homeowners not to descend into a brutalistic mindset. Rationality may demand beauty without betraying itself.

Everyone's situation is different. No one knows you and your family better than you and your family. You are autonomous. Your feelings matter. I know perfection is impossible. I believe healing is achievable. Godspeed.

References

[1] Curriculum Research & Development Group. *Practices of Science: Interpreting Safety Information.* Part of the Exploring Our Fluid Earth curriculum. University of Hawai i at Mānoa. 2011. URL: https://manoa.hawaii.edu/exploringourfluidearth/chemical/matter/properties-matter/practices-science-interpreting-safety-information (visited on 12/23/2025).

[2] Neil Nathan. *Toxic: Heal Your Body from Mold Toxicity, Lyme Disease, Multiple Chemical Sensitivities, and Chronic Environmental Illness.* 2nd. Paperback. Victory Belt Publishing, 2025. ISBN: 978-1-62860-566-2.

Chapter 12

Mold

Throughout this book, I have made numerous recommendations in service of the prevention of mold growth. This brief chapter will outline my philosophy on mold. I will touch on detection, prevalence, prevention, remediation, sensitivities, and individual susceptibility.

Few people need to be convinced that household mold is dangerous. Our biological programming tells us that it is disgusting. We feel a natural urge to get away from it. Still, much disagreement can be had about the nature and scale of the problem. The following key facts inform our judgement and recommendations.

12.1 Prevalence

Estimates suggest that indoor dampness and mold occurs in approximately 47% of North American homes with some studies suggesting even higher prevalence.[7] Mold spores are omnipresent in the natural outdoor air. Dangerous mold growth requires isolated indoor conditions where individual species can grow without interference from other species.[5]

12.2 Susceptibility

Approximately 25% of the population is suspected to be genetically susceptible to mold toxicity. It is likely that anyone can become susceptible with sufficiently high exposure. Genetically susceptible individuals are often the only members of their families that exhibit symptoms.[5]

According to publicly available research, exposure to mold is associated with increases in adverse health outcomes including asthma and other respiratory conditions. Exposure is also associated with a host of non-respiratory effects, including carcinogenic, neurological, immunologic, cognitive, ophthalmologic, dermatologic, gastrointestinal, and multi-system effects.[7] According to functional medicine doctors specializing in mold toxicity, mold toxicity is often the root cause of Chronic Inflammatory Response Syndrome (CIRS), Multiple Chemical Sensitivity (MCS), Electromagnetic Hypersensitivity (EHS), and Mast Cell Activation Syndrome (MCAS).[5]

12.3 Prevention

The optimal strategy to prevent mold growth is to eliminate the conditions of growth. Airborne mold spores are easily filterable due to their large size relative to other pathogens (approximately 10^{-6} to 10^{-4} meters). Mold generally requires a moist and digestible substrate to grow upon. The best strategies for the prevention of mold growth are (1) moisture control and (2) the use of non-digestible building materials. Typically, relative humidity levels below 50% will prevent the growth of mold.

12.4 Remediation

While mold is a living organism, it is best to approach mold as though it cannot be killed. Its various parts (mycelium, hyphae, and spores) behave differently and affect people differently, whether alive or dead. Mold often causes harm through the emission of mycotoxins that are inhaled by occupants through dust particles. Mycotoxins cannot be killed.[5] Certain antifungals are suspected to stimulate the release of mycotoxins.[6,4]

Mold can protect itself from antifungals by excreting biofilms. Biofilms are gelatinous barriers generated by fungi that buffer the penetration of antifungals. These biofilms seek to protect the central mycelium so that it can survive and regenerate.[5] The remediation strategy should not change based on whether the mold is suspected to be alive or dead.

A competent remediator is likely to follow the IICRC S520 protocol, which emphasizes physical removal via replacement, sanding, or scraping.[3] Total mold removal, while possible, is not always achievable for homeowners. Sophisticated remediation solutions for complex mold problems are very labor-intensive. They can cost up to $50 per square foot of affected indoor area.[2] (The average U.S. home would cost over $100,000 to thoroughly remediate through a sophisticated vendor.)

As always, our goal as reasonable homeowners should be to work within our means to bring occupant exposure down to a tolerable level. One alternative to total physical removal is a specialized limewash product called Caliwel. Caliwel uses an engineered calcium hydroxide solution to raise the pH of the substrate to about 12.5, rendering it inhospitable to mold. The pH level stays elevated for a minimum of 5 years after application. Lime is not a fungicide in a traditional sense. It a broad-spectrum biocide that works by altering the conditions of the substrate on which the mold grows. It does not encourage the release of mycotoxins like other fungicides. Spore and mycotoxin release during treatment is equivalent to other non-fungicidal methods of removal. But, because this process eliminates the need to sand and scrub of the substrate, it significantly reduces of the labor cost of remediation.

12.5 Detection

Testing for the presence of mold is a notoriously thorny problem.[1] There are many methods out there. Many methods test for things that are not directly relevant to occupant health.

First, let us address the easy case. In my professional experience, visible mold does not need to be tested. It merely needs to be removed. Now, let us address invisible mold.

To detect invisible mold (e.g. under floors, behind walls, or in

HVAC equipment) I recommend using a microbial VOC ("mVOC") test. In my professional experience, the airborne mVOC test provides the best representation of the mold-related health risks in a home managed by mold-aware occupants. My theory is as follows.

mVOCs are gases emitted by the active growth of mold (and some other potentially harmful microorganisms). Given that a homeowner takes the following preliminary steps, the airborne mVOC test will cover the remaining outstanding concerns. First, the homeowner (with professional assistance) should investigate and remediate all visibly water-damaged areas of the home. Then, the homeowner should adopt regular cleaning practices to reduce the accumulation of dust particles that transport mold-related toxins. Finally, the homeowner should install sufficient continuous filtration to capture airborne mold spores. After the homeowner completes these basic steps, the only remaining concern is the possibility of unidentified active growth. The mVOC test specifically addresses this concern. If an mVOC test returns a positive result within this context, further investigation and remediation is warranted. Homeowners are urged to consult with a reliable professional and respect license laws in applicable states.

References

[1] Kathleen Hess-Kosa. *Indoor Air Quality: The Latest Sampling and Analytical Methods.* CRC Press, 2019.

[2] HomeCleanse. *HomeCleanse Services.* Services page of HomeCleanse, an exclusive mold remediation and home wellness company led by Michael Rubino. Accessed December 23, 2025. 2025. URL: https://www.homecleanse.com/home-cleanse-services/ (visited on 12/23/2025).

[3] Institute of Inspection, Cleaning and Restoration Certification (IICRC). *ANSI/IICRC S520: Standard for Professional Mold Remediation.* Tech. rep. Fourth Edition (2024). ANSI-approved procedural standard for mold remediation in structures and contents. Published by IICRC and available on their webstore. Institute of Inspection, Cleaning and Restoration Certification, 2024. URL: https://iicrc.org/s520/ (visited on 12/23/2025).

[4] Chao Ju et al. "Effects of Fungicides and Nontarget Pesticides on Accumulation of the Mycotoxin Deoxynivlenol in Wheat". In: *Toxics* 11.9 (Sept. 2023), p. 768. ISSN: 2305-6304. DOI: 10.3390/toxics11090768.

[5] Neil Nathan. *Toxic: Heal Your Body from Mold Toxicity, Lyme Disease, Multiple Chemical Sensitivities, and Chronic Environmental Illness.* 2nd. Paperback. Victory Belt Publishing, 2025. ISBN: 978-1-62860-566-2.

[6] Nadia Ponts. "Mycotoxins are a component of Fusarium graminearum stress-response system". In: *Frontiers in Microbiology* 6 (Nov. 2015). ISSN: 1664-302X. DOI: 10.3389/fmicb.2015.01234.

[7] *Promoting Policy Solutions for Household Dampness and Mold: Strategies for Protecting Our Health.* Change The Air Foundation. URL: https://changetheairfoundation.org/wp-content/uploads/2025/01/CtAPolicyBrief.pdf.

Chapter 13

Case Studies

I wish that breathing healthful indoor air was a straightforward exercise. I wish that there existed a simple checklist of best practices that could all-but-ensure a healthy indoor environment. Unfortunately, this is not the case. All buildings are living machines — machines requiring thoughtful engineering. All engineering requires curiosity and critical thinking. These case studies will attempt to train readers to spot anomolies and demonstrate the steps to successful remediation. The stories within are fictional simplifications of real events.

13.1 Not-so-fresh air intake

Facts A mother and her two children in the northeast move into a newly built home. The home is in a densely packed and posh HOA (homeowners association). Shortly thereafter, the kids start developing dry scratchy throats and a persistent cough.

Investigation The mother took the kids to the doctor. The doctor examined the kids and diagnosed them with acute inhalation toxicity. No details were discernable as to what the children were inhaling that was causing their symptoms. The mother hired an IEP (indoor environmental professional) to investigate the house. The IEP discovered that there was a fresh air intake running on the back of the first floor of the house. The IEP asked the mother some questions to assess whether or not the fresh air intake could be pulling toxins into the home. (A fresh air intake is like a primitive one-way ERV. It pulls air in but does not exhaust or condition air.) The mother shared that the HOA is responsible for maintaing the lawns in the neighborhood, and she frequently observed the HOA landscapers spraying her backyard. Upon inquiry with the HOA, the IEP determined that the HOA was spraying broad-spectrum herbicides into the backyard. The IEP concluded that a significant quantity of these herbicides were being sucked into the house via the fresh air intake. The IEP and the mother agreed that this hypothesis was strong enough to take action against.

Resolution The mother hired an HVAC contractor to uninstall the ground-floor fresh air intake and replace it with an ERV vented through the roof. The mother installed standalone air purifiers (with an activated carbon layer) in the children's bedrooms to address lingering pollution. The children's symptoms resolved with time.

Lesson This investigation discovered a clear culprit — the fresh air intake and the herbicides. The amount of effort required to identify this culprit cannot be understated. Further, the human sufferring that precipitated during the investigation cannot be understated. Still, the mother in the story followed a logical course of action. As a result, she resolved the previously unidentified problem in about as quick a time as anyone could. The most important step the mother took was taking her children to the doctor. The family was very fortunate that the doctor was able to easily identify the symptoms as being associated with acute inhalation toxicity. This is a difficult call for a doctor to make. It requires a lot of experience and professional judgement. But, that call was critical to putting the mother on the track to discovering the problem. Imagine, for example, if a different medical professional assigned a non-environmental cause to the children's symptoms. The timeline to resolution could have been much different.

13.2 Crawlspace encapsulation nightmare

Facts A couple purchased an existing home. They began renovating the home in preparation to move in and live there full time. They planned to complete certain major renovations then move into the home while other minor renovations continued. First, they encapsulated the crawlspace. They made sure to also install a crawlspace dehumidifier and a sump pump. (A sump pump will automatically move water away from the foundation to prevent moisture intrusion and flooding.) After the crawlspace was finished, the couple moved into the home. They lived in one half of the home while the other half was being renovated. They began to feel violently ill with flu-like symptoms all the time. They had always noticed a chemically smell from the one half of the house that was still being renovated. They disregarded the smell, assuming it was a temporary issue related to the renovations. On a planned vacation to their extended family, their symptoms spontaneously resolved. They began to suspect their house was harming them.

Investigation They began running some common indoor air quality tests to no avail. They proceeded with renovations in the meantime. One renovation project involved replacing the carpet in one half of the house. They discovered a trapdoor under the carpet. At this point, given the previously unknown trapdoor and a handful of negative tests, they began to suspect the crawlspace was the culprit. They ordered some highly specialized tests designed to detect soil contaminants. The tests discovered a banned pesticide called chlordane in the soil. [TODO: Check the name.] They concluded that the trenching for the drainlines of the sump pump overturned old deposits of chlordane. They also discovered imperfections in the crawlspace encapsulation near the sump pump. Worse yet, they discovered that significant negative relative indoor air pressure was accelerating the uptake of chlordane into the living space. They hypothesized that chlordane was traveling from the ground around the sump pump, around the vapor barrier, through the trapdoor, and into the living space.

Resolution First, the couple sealed the trapdoor. Then, the couple had the encapsulation reinstalled using a specialized vapor barrier designed to block underground chemical emmissions. They made sure to have the encapsulation gaps near the sump pump trenching fortified along the way. Knowing they would be living atop a chlordane deposit, they installed an exhaust fan in their crawlspace to be safe.

Lessons The couple in this story was incredibly unlucky. In any case, the problem has to be identified before it can be fixed.

The couple's vacation to their extended family allowed them to identify the problem. If you expect you are suffering from acute toxicity, please try taking a vacation. Clearly identifiable medical symptoms that improve when away from the home are valuable data for you, your doctor, and any IEPs that will ultimately assist you. An infection travels with you, but a toxin does not. Merely spending a few days away from the house can help you make that important distinction.

13.3 Bleach and paint

Facts A couple recently purchased their dream home. They had ample time and energy to start DIYing some renovations. They ripped up the tile in a guest bathroom, attempting an aesthetic upgrade, and found black growth on the OSB subfloor. They researched on the internet how they should handle the growth. Following the advice they read online, they purchased some household bleach and some antifungal paint . They applied the bleach and let it dry, scrubbing a little bit of the growth away in the process. Then, they applied the paint and let it dry. They proceeded with their bathroom remodel unperturbed. They took their time DIYing the bathroom remodel as a sort of hobby project. It would be six months until they covered that particular section of subfloor back up with new tile. Twelve months into owning the house, they were sufferring from the quintessential symptoms of mold toxicity, including general malaise, depression, anxiety, rashes, and respiratory irritation.

Investigation Upon presenting their symptoms to a doctor, he recommended indoor air quality testing to attempt to find a diagnosis. They met with an IEP who asked them some questions about their house. His investigation was brief and straightforward. The IEP cut out some sections of drywall and discovered that mold had taken hold in most of the framing of the house.

Resolution The couple was forced to move out temporarily to put the house through intense remediation. They spent nearly $200,000 to save the house. All of the drywall was taken off of the frame, and then the mold-affected framing was ground down with special tools.

Lesson This story illustrates the dangers of incorrectly handling mold. It is likely that the couple turned a localized mold problem into a catastrophic mold problem through mismanagement. Publicly available information about handling mold is often dangerous and outdated. The "bleach and paint" approach is known as the kiss of death in the mold remediation industry. Adding these sorts of chemicals to actively growing mold can exacerbate the development of mycotoxins and biofilms, making the mold both more hazardous and harder to kill in the future. This is why we suggest that mold is treated as though it is not killable but only removable. The couple also failed to practice the basics of encapsulation. By leaving the mold directly exposed to the indoor air, they allowed it to spread spores to other parts of the house.

13.4 Unexplainable rash

Facts A mother with MCS developed a mysterious rash on most of her body, almost overnight. She had been living without any notable symptoms for the few months leading up to the development of the rash. Her house had recently undergone miscellaneous renovations intended to make it more tolerable for her. The mother suspected that one of the recent renovations unintentionally introduced a new source of indoor air pollution. She performed numerous indoor air quality tests. She discovered a few mild sources of indoor air pollution. Some were tangential to the renovations, but others were pre-existing. She aggressively remediated all of the problems she discovered. Her rash persisted and worsened. Feeling out of control, she booked an appointment with her doctor.

Investigation At the appointment, the doctor diagnosed her with contact dermatitis. The doctor identified that the rash was concentrated along chaffing points of normal clothing. Based on the appearance of the rash itself, the doctor ruled out chronic conditions like eczema and acute conditions like inhalation toxicity (both conditions the mother suspected). The doctor inquired with the mother about any recent changes in her clothing brands or laundering habits. The mother recalled that she recently inherited a batch of hand-me-downs for her children from a neighboring family. She recalls laundering all of the hand-me-downs in her favorite detergent in preparation to share the clothes with her children. Even after laundering the hand-me-downs, she could still smell the offensive odor of common dryer sheets (which is notoriously harmful to MCS patients) on the pieces. She ultimately decided to donate the clothes to charity. The mother hypothesized that the hand-me-downs polluted her washing machine. She suspected that the dryer sheet residue seeped into her clothes and caused her rash.

Resolution The mother tested her hypothesis by wearing some old clothes that had not yet been laundered in the polluted machine. Her symptoms improved immediately and her rash resolved within a few days. She ultimately decided to replace the washing machine. And, she may have to get rid of some of the affected clothing.

Lesson The problem was not the house this time. (We will refrain from commenting on housekeeping practices in this book. It suffices to say that MCS patients are notoriously sensitive to the fumes and residues of common dryer sheets.) This case study is the practical opposite of the above case study entitled "Not-so-fresh air intake". This case study demonstrates the anguish of incorrectly attributing medical conditions to indoor air quality problems. One of the most helpful things a doctor can do is to pattern-match the symptoms of the patient to other cases he has seen in his career. In the other case study, the doctor was able to easily identify that the children were suffering from inhalation toxicity. The doctor did so not by knowing what the children were breathing but by pattern-matching the symptoms of inhalation toxicity to similar cases. In this case study, the doctor did not know what was irritating the skin. Still, the doctor could say with near certainty that the toxin was being applied to the skin (i.e. dermal toxicity) rather than bubbling up through the skin (i.e. inhalation toxicity). Prior to her appointment, the mother was remediating indoor air pollution in the home in an attempt to treat her rash. In doing so, she had impliedly diagnosed herself with inhalation toxicity. Her remediations improved her home but not her symptoms. This type of problem is fairly common, and the emotional toll can be severe. The mother might feel foolish for misdirecting her time, energy, and financial resources. Or, she might feel grateful for the inspiration to further protect her loved ones from indoor air pollution.

Index

www.ingramcontent.com/pod-product-compliance
Lightning Source LLC
LaVergne TN
LVHW022217200926
848481LV00006B/2007
9798234090546